T0295639

Disconnected

THE WORKING CLASS
IN AMERICAN HISTORY

For a list of books in the series, please see our website at www.press.uillinois.edu.

Disconnected

Call Center Workers Fight for Good Jobs in the Digital Age

DEBBIE J. GOLDMAN

UNIVERSITY OF ILLINOIS PRESS
Urbana, Chicago, and Springfield

1 2 3 4 5 C P 5 4 3 2 1

♾ This book is printed on acid-free paper.

Library of Congress Cataloging-in-Publication Data

Names: Goldman, Debbie, author.
Title: Disconnected : call center workers fight for good jobs in the digital age / Debbie J. Goldman.
Description: Urbana : University of Illinois Press, [2024] | Series: The working class in American history | Includes bibliographical references and index.
Identifiers: LCCN 2024005970 (print) | LCCN 2024005971 (ebook) | ISBN 9780252046056 (cloth) | ISBN 9780252088155 (paperback) | ISBN 9780252047237 (ebook)
Subjects: LCSH: Call center agents. | Telecommunication. | Working class.
Classification: LCC HE8788 .G65 2024 (print) | LCC HE8788 (ebook) | DDC 331.7/6138464—dc23/eng/20240412
LC record available at https://lccn.loc.gov/2024005970
LC ebook record available at https://lccn.loc.gov/2024005971

Contents

Preface and Acknowledgments

I will always remember my first visit to a Bell system call center. It was 1992, not long after I began work in the Communications Workers of America (CWA) research department. George Kohl, the research director, had given me an assignment to become an expert in the issues facing customer service representatives, the fastest-growing occupational group in the telephone companies that we represented. Kohl arranged for Hazel Dellavia to take me on a site visit to a call center operated by New Jersey Bell, a subsidiary of Bell Atlantic. (Bell Atlantic has since merged into Verizon Communications.) Dellavia was then a CWA staff representative, having risen through the ranks as a New Jersey Bell operator and service representative to become president of her predominantly female local union.

At the New Jersey Bell call center, we stood with the manager in his office, our eyes fixed on his computer screen. He proudly demonstrated the latest technology recently introduced into the company's call centers, an automatic call distributor (ACD) programmed not only to send incoming calls to the next available representative, but also to track, record, and make visual in real time and through detailed printouts how the service representatives spent their time and did their work. The manager gleefully toggled between several screens. The first displayed multiple boxes, each representing one service representative in the office, color-coded to indicate whether she was on a call. Then the manager clicked on one of the boxes, zooming in to reveal the work activity of a particular service representative. He clicked again, this time zooming out to another color-coded screen, with each box representing a different New Jersey Bell call center. The programmed ACD allowed him, and all New Jersey Bell call center managers, to monitor, measure, compare, and benchmark activity in each call center and for each individual service representative, networked together in the statewide call queue.

The manager was delighted with the new system. Not only could he see what was going on in his call center from the comfort of his office, but also, he could generate reports for each representative and the entire office on a range of measures, including average talk time, time between calls, sales results, and adherence to the daily schedule. The company could use this data to track the volume of calls at different time periods, squeeze out downtime, and then schedule representatives to meet the exact workload. The call center manager was thrilled with the improvements in productivity and sales that the enhanced ACD enabled. Dellavia and I were appalled, anticipating the ways management would use the data to intensify surveillance, speed up the pace of work, increase sales quotas, and enhance control over the way the service representative did her job. Dellavia, who had worked as a service representative in the days of paper records and personal supervision, saw a stress epidemic in the making.

This was my introduction to the legacy Bell system digital workplace. Over the years, I would visit many other AT&T and Bell Atlantic call centers, sitting side by side with representatives taking calls and listening to managers brief me on the automated systems. My last visit was in 2014. On that visit, the call center manager demonstrated the reporting capabilities of the much-enhanced ACD. As I recall, he could generate reports showing each representative's performance on about three dozen different productivity and sales performance measures. He could record calls and review all activity at a later time.

Dellavia was right. The manner in which the Bell employers deployed the digital technologies did indeed create stress epidemics in the call centers. The customer service representatives demanded that their union mobilize to win protections against the debilitating and dehumanizing management systems that the electronic systems enabled. During the twenty-eight years (1992–2020) that I worked in the CWA research department, I assisted union leaders in contract negotiations and enforcement, organizing, labor-management initiatives, and work stoppages designed to improve conditions in the call centers and to give the service representatives greater control over the way they did their work. In their struggles, they joined a long tradition in the battle between labor and capital over control on the job and in the larger society.

During my tenure in the CWA research department, I also enrolled in a graduate program in labor history at the University of Maryland. I wanted to gain a deeper understanding of the forces in the political economy driving the changes that shaped the terrain on which CWA members—and indeed all working people—contended with their employers for good, humane jobs. I wrote the initial version of this book as my PhD dissertation, drawing on academic literature, union records, and dozens of interviews with women and men who worked in the AT&T and Bell Atlantic call centers. I hungered for access to internal company documents to understand the complex pressures behind key corporate decisions. They are not available.

Many of the chapters in this study draw on my career as a member (and in the latter years, director) of the CWA research department. I supported the annual customer service conferences described in chapter 2 and served as an expert witness in the US West reasonable sales objectives arbitration highlighted in that same chapter. I provided research to CWA organizers at Sprint and Southwestern Bell Mobile Systems discussed in chapter 3, tracked progress of union-management job redesign projects featured in chapter 4, analyzed data for the union members of the AT&T/CWA outsourcing committee described in chapter 5, and facilitated development of the service representative bargaining agenda that was a key objective of the Verizon strike examined in chapter 6. After passage of the Telecommunications Act of 1996, I directed CWA's telecommunications policy program, with multiple interventions before the Federal Communications Commission, state regulatory commissions, the U.S. Department of Justice, and Congress.

My personal background and "insider" status influence this study in several ways. First, my female gender, my union status, and my relationship with many of the CWA call center and other union leaders opened many doors, built and sustained trust and rapport in extensive interviews, provided access to key documents, helped me identify important struggles that informed my case studies, and facilitated introductions to call center workers and other union and management leaders that were essential to my research. On the other hand, my union status likely closed the door to several requests to former senior-level AT&T managers who declined a formal interview; it also likely played a role in the AT&T legal department's decision to refuse access to certain documents provided under confidentiality to the union.

Second, my insider status and background give me deeper understanding of the world of the service representative, her struggles for good jobs and meaningful work, internal union political dynamics, and the impact of regulatory change. My deep knowledge of the call center workforce and the telecommunications industry helped me ask the right questions and probe more deeply in analyzing the impact of social, economic, and political forces on the frontline workforce. My study would certainly have benefited from access to company managers, executives, and documents that described the options and rationale behind key management decisions with the same level of detail as the union documents.

I have written *Disconnected* with a clear point of view. I believe unions empower workers, capital concedes only what labor power successfully demands, and government institutions profoundly affect the outcome of the contest between labor and capital in society and in the workplace. My goal in this project is to take advantage of this insider background to present a rich, nuanced portrait of the ways call center workers experienced the impact of technological change, neoliberalism, and financialization on the job, how they fought against the degradation of conditions at work, and the constraints they faced in the context of the larger political economy.

This project has been a long time coming, and many people have helped me. I want to acknowledge the CWA leaders, members, and staff who taught me about the world of the Bell system service representatives. The CWA leaders who built the customer service network are a remarkable, creative, feisty group of women who know how to fight for power within their union and to fight their employers on behalf of their members. I particularly thank Hazel Dellavia, Mary Ellen Mazzeo, Sandy Kmetyk, Gail Evans, Annie Hill, Laura Unger, Mary Lou Schaffer, Victoria Kinzer, Mary Ann Alt, Colleen Downing, Linda Kramer, and Mary Jo Reilly. Among the CWA organizers, Sandy Rusher and Danny Fetonte (of blessed memory) answered all my questions. Nell Geiser and Dan Reynolds in the CWA research department responded to my many inquiries and continue to provide brilliant, dedicated support to advance the union's bargaining, organizing, and legislative program. Former CWA Presidents Morton Bahr, Larry Cohen, and Chris Shelton, and my personal mentor George Kohl, provided inspiration, insight, and passion to guide my work at CWA and in writing this book. Finally, I thank the former AT&T and Bell Atlantic managers who helped me understand the competitive pressures that shaped company decisions: William Stake, Steven Leonard, Mike Kzirian, and Michelle Guckert.

The University of Maryland history department is a special place. My teachers encouraged me as I combined work with academic studies. Gary Gerstle set me on the path for this book in my first research paper on the CWA operators' campaign for job control discussed in chapter 1. Julie Greene, my dissertation advisor, mentored me with probing questions, editorial guidance, and loving support, continually prompting me to "complicate" my story. I am grateful to my dissertation committee, David Sicilia, Robyn Muncy, and Meredith A. Kleykamp, for their comments and encouragement, and especially acknowledge Joseph McCartin's detailed line observations and enthusiasm for this project. Nelson Lichtenstein's support got me the book contract, and his careful early read of the manuscript challenged me to think more deeply, particularly to answer his critique of my chapter on union-management initiatives. Kyle Pruitt was my dissertation buddy extraordinaire, and Jodi Hall facilitated many administrative requests with efficiency and good cheer. Jason Resnikoff, fellow scholar of technology in the workplace, provided valuable comments, as did an anonymous reader who helped me situate my study within the literature on labor feminism. Leon Fink and two anonymous editors for the journal *Labor and Working Class History* sharpened my argument on organizing in chapter 3.

As I revised the manuscript for book publication, I was fortunate to receive valuable comments on chapter drafts from participants in the Washington, DC, labor history seminar, the Newberry Library labor history scholarly seminar, John Beck's Michigan State University labor history brown bag, the Cornell University School of Industrial Relations weekly seminar, and presentations at the Labor and Working Class History Association and Labor and Industrial Relations Association conferences. A special thank you to commentators Lane Windham, Michael Stamm, and Marcia Walker-McWilliams. Working with staff at the University of Illinois Press has been a special pleasure. Alison Syring enthusiastically championed my manuscript through the editorial process, with assistance from Leigh Ann Cowan, Megan Donnan, Allison Torres Burtka, Judy Davis, Kevin Cunningham, and Heather Gernenz.

I am fortunate to have scholar friends who provided invaluable support: Rosemary Batt, Eileen Appelbaum, Claire Goldstene, Phillip Brenner, and Virginia Doellgast. Claudia Townsend and Louise Millikan helped with editing at crucial moments. Finally, thank you to my family and friends who have listened to endless stories about the CWA call center workers and their fight for dignity, respect, and humane conditions: my sons Ben and Josh Healey, my daughters-in-law Esther Healey and Danielle Holly, my siblings Fran, Amy, and Rick Goldman, my cousin Marty Gartzman, and many dear friends but especially Betsy Vieth, Kathy Lazarus, and Louise Novotny.

Abbreviations

AFL-CIO	American Federation of Labor-Congress of Industrial Organizations
ALJ	Administrative Law Judge
AWT	Average Work Time
AT&T	American Telephone and Telegraph
BCS	Business Communications Services
BLS	Bureau of Labor Statistics
BOC	Bell Operating Company
C&P	Chesapeake and Potomac Telephone Company
CWA	Communications Workers of America
CWAC	Concerned Women's Advancement Committee
CSSS	Customer Sales and Service Specialist
CWA-TL	Communications Workers of America Tamiment Library Collection
EDS	Electronic Data Systems
EEOC	Equal Employment Opportunity Commission
FCC	Federal Communications Commission
FWTW	Federation of Women Telephone Workers
IBEW	International Brotherhood of Electrical Workers
IVR	Interactive Voice Response System
LCF	La Conexion Familiar, a Sprint subsidiary
NAFTA	North America Free Trade Agreement
NCR	National Cash Register Corporation
NLRA	National Labor Relations Act
NLRB	National Labor Relations Board
NOW	National Organization of Women

NFTW	National Federation of Telephone Workers
QWL	Quality of Work Life
RBOC	Regional Bell Operating Company
SBC	Southwestern Bell Corporation
SBMS	Southwestern Bell Mobile Systems
STRM	Sindicato de Telefonistas de la Republica Mexicana
T&T	CWA Telecommunications and Technologies division
ULP	Unfair Labor Practice
WPOF	Workplace of the Future

Disconnected

Introduction

Victoria Kintzer sat at her desk in the Bell of Pennsylvania residential business office in 1980, heard the ring of the black rotary-dial telephone on her desk, and before it could ring again, she picked up the receiver. The customer wanted to place an order for telephone service and also had a question about her bill. Kintzer looked for the customer's record in the file cabinet next to her desk. The file cabinet, known as a tub, contained all the records of customers in the local telephone exchange for which she was responsible. If the customer lived in a nearby exchange, Kintzer got up to get the record from a coworker's tub, stopping for a moment to chat, and then returned to her desk to assist the customer. She wrote up the customer's request and put her notes in the outbox for the service order writer to type up for distribution to the proper department. Although her supervisor listened in on some of her calls to ensure conformance with the detailed Bell system methods and procedures, and she resented the rules that made her raise her hand to go to the bathroom, Kintzer had relative autonomy to use her skills, knowledge, and emotional intelligence to assist her customers. She took satisfaction in helping them, solving problems, and making sure the job was done right from start to finish. She knew that she could turn to her union if she had problems on the job. This was the work life of the Bell telephone service representative in the monopoly era in the 1970s and early 1980s, before the introduction of digital technologies and market forces drove change in the customer service operations.

Fast forward to 1994. Kintzer still worked for Bell of Pennsylvania, but her job as a service representative had changed dramatically. It was a decade since the 1984 breakup of the Bell system that accelerated market competition in the telecommunications industry. Kintzer now sat in a cubicle with shoulder-high dividers separating her from her coworkers. She trained her eyes on her computer, with the automatic call distributor dropping one call after another into her headset, with no time between calls. She had no opportunity to get up, walk around, or converse briefly with her coworkers. She took only sales calls; the automated system sent billing inquiries to a different call queue staffed by lower-paid collections representatives. The job had become quite complex, with various rates for different geographic areas, multiple products and services, and data entry and retrieval from several software systems. Kintzer entered her service orders directly into the computer, frequently overlapping this task as she took the next call. Electronic scripts guided her through conversations with customers, designed to upsell additional products and services. Digital software tracked her every move in real time—how long she took on each call, the time between calls, adherence to her daily schedule, sales performance, the manner in which she moved through her scripts and databases, and even how many keystrokes she took to perform a task. Supervisors evaluated her performance on all these measures and imposed discipline for failure to meet sales quotas and time measurement benchmarks. As a local union leader, Kintzer faced challenges protecting her members from the intensified monitoring, sales pressures, work speed-up, and reclassified lower-wage job titles. This was the work life of the service representatives in the Bell telephone companies in an era of increased competition and fully automated workplace management systems.

Within the space of a decade, Bell of Pennsylvania, along with the other legacy Bell system companies, had revolutionized the work life of tens of thousands of service representatives employed in the customer care call centers. As corporate executives transformed their businesses from monopoly-era regulated public service organizations with guaranteed rates of return to competitive enterprises in pursuit of ever-higher profits, the managers of the customer service operations struggled to reduce costs and boost revenues while still providing high-quality service. These managers turned to Bell system engineers, programmers, and outside vendors to enhance the digital technologies that distributed the calls to the service representatives. They adopted sophisticated workforce management tools that enabled them to intensify control and surveillance over the pace and manner in which the frontline customer service representatives did their work. The resulting work organization of high demand with little control was a recipe for debilitating and demoralizing stress on the job.[1]

How did the call center workers—and their union—wrestle with management over the degradation of working conditions and downward pressure on living standards in the call centers? And how successful were they in their struggles for good, humane jobs in the highly automated service centers? *Disconnected* addresses these questions through a historical study of a predominantly female customer service labor force at two legacy Bell companies, AT&T and Bell Atlantic (now called Verizon). I argue that the service representatives coalesced into a workforce of resistance in response to the degrading, stressful conditions in their call centers. I use the term "workforce of resistance" to emphasize collective action, rather than individual acts of sabotage or subterfuge. The service representatives mobilized as an occupational group to pressure their union, the Communications Workers of America (CWA), to fight on their behalf for humane, safe, secure, well-paying jobs. Their collective resistance took many forms and used multiple strategies. But whether through collective bargaining, contract enforcement, union-management joint projects, strike action, or organizing, their fight for greater control at work and for compensation commensurate with the value they created for their employers represented acts of resistance.

Their struggle for good jobs faced many challenges. The very forces driving change in their call centers—the introduction of new digital technologies, the intensification of market competition facilitated by those technologies, neoliberal regulatory policies, the financial turn in capitalism and corporate governance, and the decline of unions—also narrowed the terrain on which these workers struggled with management to shape their conditions at work. The customer service workers served as the shock troops on the frontlines of this tumultuous transition, and they often felt like cannon fodder.

Disconnected argues that while deployment of new technology enables the reconfiguration of the economic terrain, it is human beings, with competing interests, resources, and power, who make the decisions about how to deploy technology and which institutional structures will frame the social impact of those technologies. The decisions are social and political choices, with outcomes determined by political struggles between management and labor, shaped by the institutional structures of the political economy in which they take place. As Karl Marx wrote, "People make their own history, but they do not make it just as they please; they do not make it under circumstances chosen by themselves, but under circumstances directly found, given, and transmitted from the past."[2]

In their battles with management, the AT&T and Bell Atlantic customer service workers had one important resource that most U.S. call center workers, indeed most service-sector employees, do not have: they were represented by a union. The CWA had a mature bargaining relationship with the Bell companies dating back to the post–World War II period, giving them a vehicle for collective resistance not available to most call center employees. To be sure, individual call center workers have some degree of agency in pushing back against conditions in the digital workplace—they can game the system, they can quit—but without organization, they have limited power to influence fundamental issues of work organization, technology deployment, job security, and compensation.[3]

The CWA-represented customer service workers struggled for power not only with their employers but also within their union. Although militant women operators built CWA, the male technicians were the dominant force within the union by the 1960s. With automated equipment gradually decimating the operator workforce, and competition elevating the size and strategic importance of the customer sales and service operation, the service representatives emerged as the female work group fighting for greater voice within the union. Labor historians have catalogued the struggles of women unionists for power within their unions, making the case that craft- and gender-based structures gave unity, voice, and strength to women unionists that women in amalgamated industrial unions battled to achieve.[4] My narrative complicates this argument, finding that while female-led customer service bargaining units excelled at crafting local solutions to workplace problems and developing strong leaders, they needed to mobilize the full power of their diversified union to make significant progress on their issues. This was a formidable task, but when successful, as during the Verizon strike in the year 2000, that unity proved critical to winning a pathbreaking stress relief package after eighteen days on the picket lines.

The gender division of labor is prominent in the call center environment, as it is in other "caring" professions. Women make up the vast majority of nurses, home care workers, flight attendants, clerical workers, wait staff, retail clerks, social workers, and elementary and high school teachers. Their daily work requires what scholar Arlie Hochschild identifies as emotional labor. Emotional labor is hard work, though frequently underappreciated and underpaid. Emotional labor can be deeply satisfying, but it can also be enormously frustrating and stressful, particularly when emotional laborers do not receive the autonomy, respect, compensation, and conditions at work that allow them to provide good service to patients, students, clients, and customers. Scholars of twentieth-century women service workers have demonstrated that female employees deployed their relationships with customers and clients, as well as their craft identities, as resources to build unity and power.[5] The Bell system call center workers also attempted to use customer relationships and an emphasis on high-quality service as sources of power and solidarity, but the machine-paced, highly scripted, and heavily monitored call center environment imposed limitations on these strategies, as did management focus on cost cutting over service quality. Moreover, forging labor-consumer alliances proved difficult as consumer advocates as well as policymakers prioritized competition and regulatory policies designed to lower consumer prices. This drove a race to the bottom in labor standards and customer service, as the unionized Bell companies strove to compete with the lower-wage, anti-union new entrants into the telecommunications market. An unintended consequence of consumer advocates' and regulators' laser focus on price competition, then, was deteriorating customer service, which has continued to this day as unionized call centers compete with a low-wage, high turnover, nonunion, and increasingly offshore workforce.

The call center activists you will meet in this book were part of the feminization of the working class and labor movement that began during the latter decades of the twentieth century. Of course, women have always worked in the paid labor force and been union members and leaders. Yet, due to structural changes in the U.S. economy, as well as cultural shifts in society, working-class women were entering the labor force in record numbers during these years. Influenced by the women's movement, the civil rights movement, and the legal protections of the Civil Rights Act of 1964, this was a dynamic period as women and people of color in nonunion locations mobilized to form unions and demand their rights on the job. *Disconnected* contributes to the growing body of labor history that challenges the all-too-dominant narrative of working-class declension in power beginning in the latter third of the twentieth century, largely based on the decline of the heavily unionized, predominantly male and white industrial sector. By widening the lens to bring working women and people of color into the picture, we see a more complex portrait of a reformulated working class organizing for equity and economic justice.[6]

As historian Katherine Turk notes, "these were years of transformative potential and missed opportunity."[7] Aggressive mobilization by business leaders and their conservative allies blocked or narrowed radical demands for greater power and equity at work and in society. Virulent anti-union employers thwarted union organizing drives. Conservative courts weakened the reach of Title IX of the Civil Rights Act. Neoliberal regulatory policies drove a race to the bottom that reversed or stymied gains for workers in formerly regulated infrastructure industries, including flight attendants and, as we shall see, telephone workers. For many, a narrow, individualistic "rights" consciousness eclipsed a more solidaristic cross-class alliance for equity and economic justice.[8] But the CWA women you will meet in this book already had their union, and they directed their time, energy, and leadership to fight for equity and dignity within and through their union. They fought simultaneously for comparable worth for their profession and affirmative action access to higher-paid technician and customer service jobs. They saw no disjuncture between the fight for equal rights and for labor rights.

Disconnected focuses on the contest between labor and capital in a pivotal industry of the post-industrial economy—telecommunications—and within that sector, customer sales and service. The digital revolution, coupled with the consumer economy's growth, transformed the ways people buy products and solicit customer service and support in the United States (and abroad). Beginning in the 1980s and accelerating since then, call centers became the primary vehicle through which businesses (and many public agencies) interact with customers, clients, and citizens.[9] An estimated 3.7 million workers in the United States are employed in almost 40,000 domestic call centers, representing about 2 percent of the U.S. workforce.[10] More U.S. workers are employed in call centers than in the manufacture of automobiles, airplanes, machinery, and steel combined.[11] The global call center industry employs millions more.[12] The U.S. domestic call center sector generated $110.6 billion in revenue in 2022; global call center revenue was $314.5 billion that same year. In 2022, the four largest worldwide call center companies employed almost 1 million workers in more than 900 global call centers.[13]

While call center employees work in virtually every sector of the U.S. and global economy, this book focuses on customer service representatives employed by two leading progenies of the Bell telephone system, AT&T and Bell Atlantic. The Bell system provides an excellent case study to unravel the multiple forces driving change and setting the bounds for organized worker resistance in the automated customer service workplaces. Two distinct regulatory events accelerated market competition in telecommunications markets. The 1984 AT&T divestiture broke up the integrated Bell system monopoly. To resolve a longstanding antitrust suit, the AT&T consent decree restricted the company to long-distance service, equipment manufacturing and installation, and research in the world-renowned Bell Labs, while requiring it to spin off the local networks to seven Regional Bell Operating Companies (RBOCs). Bell Atlantic was one of the RBOCs, operating in the mid-Atlantic region of the United States. The RBOCs were limited to providing local telephone service. Because competition came earlier to challenge AT&T's dominance in long-distance service than it did to Bell Atlantic's more protected local markets, the comparison provides another lens through which I analyze the impact of market forces and regulatory structures on the power of the union and

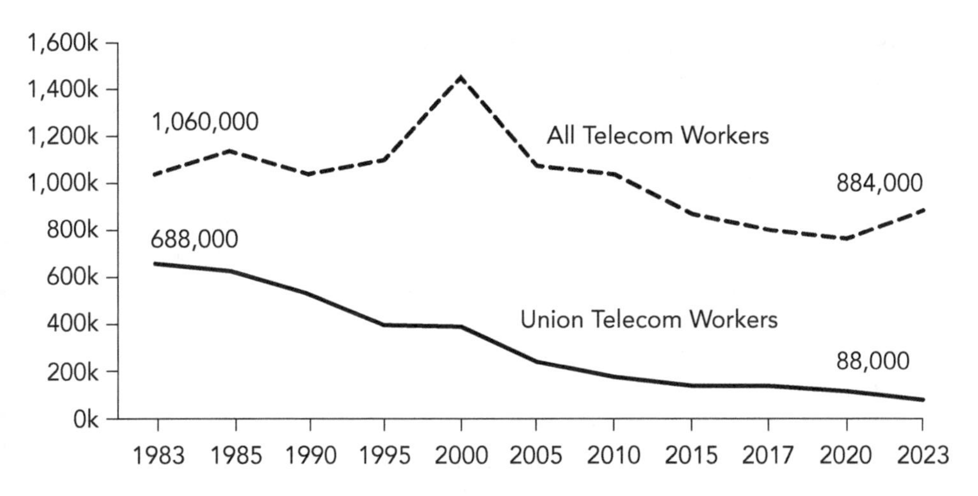

Telecommunications Employment, 1983–2023. Source: Hirsch/Macpherson database (1983–2018); Bureau of Labor Statistics (2019–2023)

call center members.[14] Bell system employees who weathered the 1984 dissolution of the AT&T monopoly talk about work life "before" and "after" divestiture. "Before" were the good old days; "after" included massive job cuts, insecurity, weakening of union power, and declining customer service.[15] Twelve years after divestiture, Congress passed the Telecommunications Act of 1996, which opened all telecommunications markets to competition and accelerated a dizzying wave of mergers, acquisitions, and cost-cutting pressures, including outsourcing of call center work. These two government policy changes serve as markers for transformation in market structure and competitive conditions, which, in turn, shifted the terrain on which the union and management contested for control. Union representation in the telecommunications industry declined during this period from about 60 percent in the early 1980s to 22 percent in 2005, the concluding year of this study. Today it stands at just under 10 percent.[16]

The call center workers waged their battles during a period of weakening union power in the United States. Union representation in the United States has dropped to just under 7 percent of the private-sector workforce.[17] Unions are essential to give call center workers, and all working people, a collective voice to counter the power of capital in the workplace and in the larger society. Government policy is critical in structuring the labor market institutions and industrial relations systems within which unions contest for power. The U.S. industrial relations system, with its decentralized enterprise-based bargaining structures, provides few, if any, formal avenues for union participation in key management decisions regarding technology deployment and work organization. In contrast, European-style works councils, codetermination policies, and industry-wide sectoral bargaining give unions greater power to influence management decisions. Sweden's Co-determination Act of 1976, for example, requires employers bound by collective agreements to consult with trade unions on all important changes in the workplace, including deployment of new technology and labor processes. Contract negotiations there take place on an industry-wide basis, reducing labor cost as a source of domestic competitive advantage. In Germany, works councils, elected by the workforce (often with union coordination), have strong codetermination and veto rights, particularly over the use of electronic monitoring, scheduling practices, design of pay for performance, health and safety, discipline, and firing decisions. In large companies, such as Germany's leading telecommunications company Deutsche Telekom, union representatives compose half the members of corporate supervisory boards.[18]

These structures make a difference for call center workers. In a global survey of call center work organization and employment practices in seventeen countries conducted in the early 2000s, researchers found that the call centers in Scandinavia, Germany, and Austria—countries with strong labor market institutions—have better jobs, lower rates of turnover, and lower wage dispersion than call centers in the more liberal market economies of the United States and Great Britain. Further research comparing U.S. and German call centers in the telecommunications industry concluded that "worker representatives (in Germany) used their strong participation rights to help managers find compromise solutions that reduced costs and improved productivity and service quality, while ensuring that the privacy, dignity, and economic interests of the workforce were respected."[19]

Although the CWA customer service union leaders and their members lacked the strong labor market institutions of their European sisters, they mobilized within the constraints of the U.S. industrial relations system to promote an alternative vision of the customer service workplace. That vision emphasized high-quality customer service delivered by highly skilled professionals working in an environment that valued their judgment, experience, and skills to respond to customer needs. CWA customer service members and their leaders mobilized to fight for working conditions in the call centers that emphasized quality rather than low cost. They believed that the benefits in customer satisfaction and loyalty, as well as the reduced costs from lower rates of turnover, absenteeism, and stress-related illness, could support higher wages, benefits, and investment in training. They articulated a program that would use the enormous power of the information generated by the digital technologies as tools to improve worker performance and customer care, rather than for monitoring, speed-up, and delivery of scripts for customer interactions.

Indeed, the union won important victories for its members: curbing secret supervisory monitoring, limiting discipline for failure to meet sales quotas and time measurement benchmarks, raising service representatives' wages, fighting the functionalization and downgrading of non-sales positions, bringing back outsourced work, and gaining access to jobs selling and servicing the internet (and at AT&T, wireless). But the union and its customer service members faced formidable challenges as their employers reorganized their businesses many times over to compete against non-union companies with lower labor costs and more favorable regulatory treatment, while at the same time meeting the demands of capital owners to deliver ever-higher returns on their investments. Weak U.S. labor laws provided little assistance to nonunion workers when they faced aggressive employers seeking to block their campaigns for union representation. The victories of the CWA call center workers frame the boundaries of organized worker power in contesting managerial control in the automated workplace in the context of the neoliberal political economy and financial turn of capitalism. The contests of this pivotal unionized workforce, operating in one of the most dynamic and important sectors of the U.S. and global economy, highlight the opportunities, challenges, and constraints U.S. service workers face in their struggles for power in the post-industrial service economy.

The digital transformation affects virtually every workplace today. From the Amazon warehouses, to UPS drivers, from tightly scheduled railroad workers to post-COVID white collar work-at-home professionals, employers deploy the vast capabilities of information technology to regulate, track, and control the pace and the manner in which employees do their jobs.[20] These electronic systems give employers tools that Frederick W. Taylor only dreamed of when he pioneered his system of scientific management in the factories of the early twentieth century. Not only do they give management the ability to control workers' time and the labor process far beyond the machine pacing of the assembly line, but they also enable employers to break up internal labor markets and to outsource production and service functions to third parties around the globe. In the call center, automated systems add a third dimension, as they mediate the interaction between workers and customers. The same digital technologies that employers use to control the labor process, technology companies use to track and monetize our lives as we move through websites, emails, texts, and social media posts. Business historian Shoshanna Zuboff calls this the *Age of Surveillance Capitalism*. This study of the ways organized workers mobilized to resist dehumanizing labor in digital workplaces, and the constraints they encountered in the context of a neoliberal political economy that provides few public-policy guardrails against the ravages of financial capitalism, has wider implications as we reexamine the regulatory regime that has made this possible.[21]

The workplace is contested terrain, an arena for struggle between capital and labor over job control and allocation of a firm's resources. To be sure, the power struggle between labor and management does not take place on a level playing field. Management owns the means of production and capital resources, and it wields the power to hire and fire. And yet, because employers, even in highly automated workplaces, depend on human beings to create value, workers retain a degree of agency over their labor power. Labor historians have created a rich literature documenting, in the words of a leading scholar, the "opaque but potent heritage of on-the-job (worker) struggles to control the terms under which they labor for a living."[22] At the turn of the industrial era, craft workers fought to preserve their "moral code" to set the level of output and manner in which they did their work. Workers at the Watertown, Massachusetts, Arsenal and elsewhere rebelled against Frederick Taylor's system of scientific management with its time motion studies, deskilled and fragmented task systems, and separation of thinking from doing. Secretaries and department store clerks defeated management's attempt to deskill and control the pace of their work. Steelworkers went on a nationwide strike to defend contract language requiring union approval of job-displacing technology. Industrial unions negotiated and enforced contractual work rules and job classifications to protect against speed-up, arbitrary and tyrannical supervisors, and job displacement. Autoworkers at the celebrated Lordstown, Ohio, General Motors plant battled the mind-numbing introduction of automated systems into their workplaces. These and countless other workers acted collectively by organizing into and strengthening their unions and walking off the job to demand greater control over their labor process and its rewards.[23]

The outcome of the contest between labor and management is not predetermined. Rather, it takes place within a specific historical context, shaped by the level of market competition, government regulatory policies, the speed and nature of technological change, the structure of the corporation, and the level of worker organization and consciousness.[24] The contest between the call center workers and their Bell employers that I describe in this volume took place in a period of major structural transformation in the U.S. (and global) political economy. These were the decades in which a free market Neoliberal Order began to eclipse the regulated capitalism of the New Deal Order. The breakup of the Bell system reflected and reinforced the emerging Neoliberal Order. Neoliberals extolled the virtues of free markets, unfettered by government regulation designed to contain the excesses and failures of market capitalism. Neoliberals supported cutbacks in public social welfare programs, in contrast to the New Deal's mildly redistributionist tax and budget policies. Neoliberals promoted a union-free environment, rejecting the New Deal Order's view of unions as a countervailing force necessary to check unbridled corporate power.[25]

Neoliberal deregulatory policies fostered the dramatic transformation of capitalism during this period, as finance, rather than industrial might, came to dominate the U.S. (and global) economy. By the early years of the twentieth century, and growing ever since, finance accounted for the largest share of the U.S. economy, contributing four times the level of profits as the manufacturing sector and the largest share of portfolio income of non-financial firms. The role of finance grew as even non-financial companies came to derive an increasing portion of revenue from financial activities, and the role of investment banks, private equity and hedge funds, and institutional investors came to exert greater influence over corporate strategies.[26]

During the years covered in this book, the former Bell companies reinvented themselves from a Chandlerian model of managerial capitalism to a financial model. Briefly stated, this transition, driven by capital's laser focus on boosting shareholder value, changed the managerial function from one that coordinated vertically integrated activities within a firm to the manipulation of assets to be bought, sold, and reengineered with the goal of boosting returns to shareholders. As free market economist Milton Friedman wrote in his famous 1970 essay, "The Social Responsibility of Business Is to Increase Its Profits," in order "to make as much money as possible" for the company's owners, as measured by the share price.[27] As a firm's financial success became less dependent on productive activity, managers increasingly viewed labor as another factor of production to be squeezed, rather than a reciprocal relationship that could add value to the company. The logical extension of the financial turn, as labor economist David Weil explains in *The Fissured Workplace*, is the outsourcing of larger and larger portions of the work producing goods and delivering services. The call center workers experienced the impact of the financial pressures on AT&T and Bell Atlantic to maximize shareholder value in the downward pressure on living standards, intensification of working conditions, and increasing job insecurity.[28]

The chapters in the book analyze the effectiveness of the multiple strategies the call center workers and their union deployed in their fight for good jobs, with particular attention to the context in which they took place and the product market serviced by the call center workers. CWA made the greatest progress when the affected customer service workforce sold higher-value products and services; when regulatory structures protected markets from downward pressure on labor costs; when their employer's financial performance was strong; when union density remained high; and when customer service members and leaders were able to mobilize the full power of the union behind their demands.

The book chapters are organized both chronologically and thematically. The first two chapters contrast the world of the customer service representative "before" and "after" the breakup of the Bell system and the introduction of digital technology into the call centers. Chapter 1, the "before" chapter, compares the relative autonomy the service representatives experienced at work with that of the highly regimented, heavily surveilled telephone operators. The chapter highlights two key developments of the pre-divestiture period: first, the landmark 1973 Equal Employment Opportunity Commission (EEOC) affirmative action consent decree, which challenged the gender division of labor in the Bell system, and second, the female operators' uprising against technology-induced job pressures. Chapter 2, the "after" chapter, describes the impact of AT&T divestiture on management decisions to intensify and degrade conditions in the call centers. In response, the service representatives mobilized as a workforce of resistance, demanding that their union negotiate contractual protections against abusive surveillance, speed-up, deskilling and downgrading of their jobs, and unreasonable sales pressures. The outcomes of the battles at AT&T and Bell Atlantic were not the same, revealing the central role that market structure, regulatory oversight, and union power played in the contest between labor and management.

The logical CWA response to downward pressure on labor standards in the competitive era was to organize the unorganized. This was easier said than done, since weak U.S. labor laws provide little protection against an aggressive employer determined to keep the union out. In chapter 3, I analyze two CWA strategic organizing campaigns, the first at Sprint Communications' long-distance operation and the second at Southwestern Bell's non-union wireless subsidiary. The Sprint campaign, which ended with call center closure and defeat, was conducted within the framework of the failed U.S. National Labor Relations Act (NLRA), while the Southwestern Bell campaign, which eventually led to the organization of 45,000 wireless workers, took place outside the NLRA electoral framework. While multiple factors were at play, the key difference was CWA's success in leveraging power at key moments of regulatory reform to neutralize Southwestern Bell's opposition, allowing wireless workers to select union representation free from fear of job loss or harassment.[29]

Throughout the 1990s, conditions in the AT&T and Bell Atlantic call centers continued to deteriorate. Responding to management's interest in labor-management partnerships to increase productivity and service quality, chapter 4 analyzes three negotiated union-management initiatives to redesign the call center job. Although CWA participants made some progress—winning a wage increase for the new job at AT&T, for example—the projects faltered as corporate executives prioritized reengineering through job cuts, outsourcing, and blocking union growth in new ventures. Yet, some of the proposals the union piloted at Bell Atlantic surfaced again as bargaining demands a few years later during the Verizon strike.

As AT&T and Bell Atlantic pursued outsourcing strategies in both core and new businesses, threatening call center workers' job security and negotiated living standards, call center leaders' attention shifted to strategies to bring outsourced work back in house. Chapter 5 contrasts the relative success at Bell Atlantic in winning jurisdiction over the growing internet business with the challenges CWA faced at the rapidly declining AT&T long-distance operation. As market share and revenue at the once-powerful AT&T dwindled, the company conceded little ground as it prepared itself for sale to SBC Communications. In contrast, Bell Atlantic continued to grow through a series of mergers, renaming itself Verizon in the year 2000, even as cost cutting intensified pressures in the call centers. Chapter 6 details the story of the 2000 Verizon strike, leading to a neutrality agreement for organizing rights at Verizon Wireless and a pathbreaking stress-relief package for the customer service representatives.

The year 2000 represented the high-water mark for the Verizon call center workers. The epilogue brings the story up to date, as Verizon violated the wireless organizing agreement, and the legendary AT&T sold itself to SBC Communications and vastly expanded call center outsourcing, leaving union representation a shell of its former self. As I complete this book, we witness a surge of worker organizing and public support for unions in this post-COVID period, including the technology and communications sectors. Whether this will lead to permanent change with the scale and scope to rebalance the struggle between capital and labor in the twenty-first century is still an open question.

CHAPTER 1

Before the Breakup

Sandy Kmetyk, dressed in white gloves and a pillbox hat, began her first day of work as a telemarketing representative in the downtown Pittsburgh office of the Bell Telephone Company of Pennsylvania in November 1967. Residents of her working-class neighborhood of North Hills, a community just outside of Pittsburgh populated by first- and second-generation Croatian and Italian immigrants, knew that the telephone company offered stable employment and decent pay, especially given the other options available to a female high-school graduate in 1967. Kmetyk was pleased when the telephone company offered her a job, and then, to her delight, gave her a promotion a month later to customer service representative in the residential business office on Pittsburgh's north side, earning a weekly wage of $81. Kmetyk would remain a telephone company service representative for the next twenty-two years, with a three-year break in the mid-1970s to care for her young children. Although Kmetyk left the company payroll in 1989, her life remained rooted in the world of Bell system customer service workers as she assumed leadership in her union, first as executive vice president and then, in 1996, president of Communications Workers of America (CWA) Local 13500, a statewide local of telephone company customer service employees throughout Pennsylvania.[1]

Kmetyk began her tenure as a customer service employee in the vast Bell Telephone system in the waning years of the New Deal Order, the set of political relationships and ideological constructions that dominated the U.S. political economy for more than four decades, from the Great Depression of the 1930s through the late 1970s and the election of Ronald Reagan in 1980. The New Deal Order, grounded in a liberal political philosophy and Keynesian economic principles, promoted government intervention in and regulation of the market to contain the excesses and failures of capitalism, fostered a moderately redistributionist social welfare state, and supported unions as a countervailing force to corporate power in the economy and polity. Throughout much of the twentieth century, the New Deal Order shaped the regulatory, managerial, and labor relations systems of the Bell Telephone system. Federal and state regulators set the rules governing the company's monopoly telephone business, insulating the corporation and its workforce from many of the economic pressures transforming business operations and labor relations in other large private-sector companies in the 1970s. The Bell system's corporate organization followed the classic bureaucratic organizational model described by business historian Alfred Chandler as managerial capitalism, with the visible hand of a hierarchy of professional managers taking the place of the invisible hand of market mechanisms in coordinating economic activity and allocation of corporate resources. Labor-management relations were firmly grounded in the collective bargaining framework of the 1935 National Labor Relations Act (NLRA), one that provided secure, relatively well-compensated jobs to more than half a million union-represented Bell employees. While the 1960s and 1970s were a time of relative stability in the Bell system and in the business offices where the customer service representatives worked, a storm was brewing as ascendant neoliberal economic policies and technological innovations opened the door to regulatory change and the 1984 AT&T divestiture that ushered in a new era in the legacy Bell telephone companies.[2]

During the monopoly era of relative market stability, the Bell companies managed the customer service employees through a highly gendered, paternalistic, and bureaucratic system. Bell company manuals described in detail the methods and procedures for almost every aspect of the job. Yet, the business offices were no assembly line. The Bell employers had not yet introduced automated systems into the customer service offices. There was no machine pacing or electronic surveillance, all transactions were recorded on paper, and the small offices fostered warm, personal relationships. Certainly, the service representatives resented many of the Bell system rules, especially those that they believed treated them as children and interfered with their independent judgment in serving their customers. At the same time, they identified with their employers' deeply ingrained culture of customer service, which gave them the opportunity to use their training, experience, and emotional intelligence to provide customers with a vital public service, that of telephone communications. During this pre-divestiture period, the service representatives lived with these contradictions. They hated the telephone companies' overbearing systems of control, yet they also found enough space within those constraints to exercise their own judgment, giving meaning and satisfaction to their work.

The service representatives were well aware of the relative autonomy they enjoyed compared to the much more numerous and lower-paid telephone operators who were heavily monitored and tightly managed by electronic technology. After the Bells introduced digital systems into the operator centers beginning in the 1960s, the women operators mobilized to demand stress relief from the increased surveillance, speed-up, and deskilling wrought by these new systems. While the service representatives were minor players in the operators' uprising, their campaign set the stage for the post-divestiture struggles of the service representatives, who would learn from and build on the operators' determined and militant stance.

Pre-Divestiture Bell System: *The* Phone Company

Throughout most of the twentieth century, American Telephone and Telegraph (AT&T) was *the* phone company, a vertically integrated, regulated monopoly that provided local telephone service to locations serving 80 percent of U.S. households and businesses through twenty-one wholly owned and two partially owned Bell Operating Companies (BOCs); long-distance and international telephony to virtually all U.S. customers through the Long Lines Department; telephone equipment to homes and businesses manufactured and distributed by its Western Electric subsidiary; and world-renowned research centered at Bell Telephone Laboratories. While the BOCs, with their own boards of directors and management teams, had a significant amount of flexibility in managing their operations, core policies were set by and directed from AT&T corporate headquarters at 195 Broadway in New York City.[3]

AT&T operated in a relatively stable environment as a regulated monopoly with little competition and a controlled pace of technological change. Policymakers considered the telephone system a natural monopoly, with efficiencies rooted in the scale and scope economies of one integrated network. The Federal Communications Commission (FCC), established by the Communications Act of 1934, regulated AT&T's long-distance service. State regulatory commissions had oversight over local and intrastate toll service. Regulators established the rates AT&T and the BOCs could charge for services with the guarantee of a "fair rate of return," usually 5 percent to 7 percent. AT&T's mission, first articulated in 1907 by then-president Theodore Vail, was "one system, one management, universal service."[4]

The service ethic permeated Bell culture. In 1971, AT&T CEO H.I. Romnes told the annual meeting of AT&T shareholders that "our first responsibility today remains what it has always been: service. . . . [I]t requires that we shun any action that is merely expedient, offering temporary advantage or momentary favor at the cost of sound long-term growth." Bell management set as its central objective during the first six decades of the twentieth century the growth and integration of the national telephone network. In 1979, fully 98 percent of U.S. households had telephone service, up from 62 percent in 1950. A grand bargain between AT&T and federal and state regulators facilitated this post–World War II expansion. In a series of agreements culminating in the Ozark Plan of 1971, the Bell system adopted below-cost pricing for local residential and rural customers cross-subsidized by above-cost pricing for business and long-distance services, providing affordable, near-universal local telephone service. The system worked as long as AT&T remained a monopoly provider to all these customer segments, but it came under stress as regulators opened the network to competition.[5]

In the late 1960s, the departments where Bell system customer service employees worked—known as the commercial and marketing departments—represented a small but growing fraction of the vast non-supervisory workforce in what was then "the biggest company on Earth." In 1967, the year Kmetyk took her job at Pennsylvania Bell, the Bell telephone system had 668,000 employees. (In addition, Western Electric had 170,000 employees and Bell Labs had another 15,000.) Eighty percent of Bell telephone workers were classified as non-supervisory, most of whom were eligible for and represented by a union. The largest occupational groups at that time were the operators (174,000), technicians (168,000), and clerks (108,000), with customer service employees (42,000) far behind. Over the course of the next fourteen years, this would change dramatically as automation allowed the Bell system to slash 73,000 operator jobs and as new network technologies, infrastructure expansion, and increased demand for services resulted in the addition of technicians, customer service employees, and clerical workers. By 1981, the last year for which the U.S. government collected comprehensive Bell system employment data, there were 677,000 Bell system non-management employees, representing three-quarters of the Bell Telephone system workforce. Among non-management employees that year, technicians (270,000) and clerical employees (155,000) dominated, while the number of customer service workers (104,000) edged out the number of operators (100,000).[6]

The Bell companies maintained a deeply gendered and race-based employment system. It was the largest employer of women in the United States. In the early 1970s, one out of every fifty-six employed U.S. women worked for the Bell system either as an operator, clerk, or service

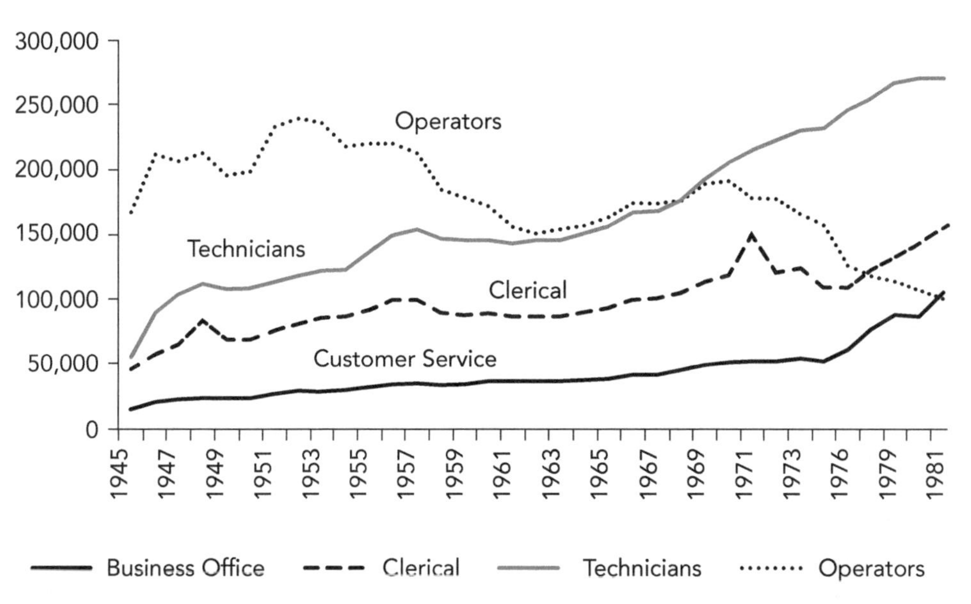

Bell System Non-Supervisory Employment by Major Occupations, 1945–1981. Source: FCC. Statistics of Common Carriers, various years, Table 11 (1945), Table 12 (1946–1950), Table 10 (1951–1981). Data for 1971–1981 extrapolated from all carrier data, discounted by 5 percent (the amount of non-Bell employment in prior years).

representative.[7] As early as the 1880s, Bell companies hired U.S.-born, white women as operators to project, in the words of historian Venus Green, a feminine image of "white ladies" as the "voice with a smile."[8] In 1967, the gender breakdown among the various job titles is striking. Among the 174,000 operators, only forty-one were men and only 821 of the 168,000 technicians were women. The Bell system carried the gender-based customer service image into the business offices. Among the 32,740 service representatives in 1970, all but 353 were women.[9] While the 1973 Equal Employment Opportunity Commission (EEOC) affirmative action consent decree opened up technician jobs to women, and operator and service representative jobs to men, these occupations remained largely gender segregated. In 1981, eight years after the affirmative action plan was implemented, a full 91 percent of operators, three-quarters of customer service employees, and only one-fifth of technicians were women.[10]

Women began their initiation into the company's gendered, paternalistic, and highly intrusive employment practices from the moment they applied for a job with Ma Bell. Before the 1970s, the Bell Operating Companies maintained separate male and female recruitment offices. In

Table 1. Bell Telephone System Non-Supervisory Employment by Gender, 1967 and 1981

	1967			1981		
	Male	Female	Female %	Male	Female	Female %
Operators	41	174,150	100%	9,051	90,036	91%
Business Office	10,683	31,279	75%	27,603	77,289	74%
Technicians	167,593	821	0%	208,973	58,910	22%
All Non-Supervisory Employees	200,341	337,385	63%	285,516	390,869	58%
All Bell Telephone Employees	290,639	377,520	57%	400,249	462,963	54%

Source: FCC. Statistics of Common Carriers, various years, Table 11 (1945), Table 12 (1946–1950), Table 10 (1951–1981). Data for 1971–1981, extrapolated from all carrier data, discounted by 5 percent.

newspaper advertisements, the companies listed operators and service representatives as female jobs and technician openings as male jobs. C&P Telephone of Washington, DC, insisted that its female applicants take a pregnancy test, ostensibly to protect against hiring a woman who would squander company training investment by leaving to have a baby. The C&P of Maryland's home visit for applicants asked the number and age of children, home duties, method of transportation, appraisal of childcare arrangements, and whether the applicant was living with her husband. An AT&T recruiting manual explained that recruiters for service representative positions should consider the plans of the applicant's husband or father and the likelihood of marriage and maternity. It encouraged recruiters to make home visits to "help establish a better understanding of the job with the family or husband of the applicant."[11] Not until 1972 did AT&T's assistant vice president for human resources, John W. Kingsbury, advise all Bell system personnel vice presidents that these types of questions and practices "might not be considered job related and of questionable utility." The EEOC case and 1973 affirmative action consent decree brought an end to these formal practices.[12]

By the late 1960s, the Bell system was no longer able to maintain a racialized white voice, because the racial composition of the operator workforce changed dramatically. In most urban areas, one-third of the operators were Black women and in heavily Hispanic El Paso and San Antonio, Texas, one-third of the operators were Hispanic. The growth of the African American and Hispanic operator labor force reflected the changing demographics among working-class women and the relatively tight labor market for women with at least a high-school education during this period. AT&T was acutely aware of this change. A 1969 report by AT&T Vice President Walter W. Straley bemoaned the fact that the tighter labor market meant that "there are not enough white middle class success-oriented men and women in the labor force—or at least that portion of the labor force available to the telephone companies—to supply our requirements for craft and occupational people." The Bell system was a vast employment engine; every year, the companies interviewed 1.9 million people and hired 200,000 of them. In a frank and racist acknowledgment of discrimination in the labor market, Straley reported that "in today's world, telephone company wages are more in line with black expectations. . . . Most of our new hires go into entry level jobs which means we must have access to an ample supply of people who will work at comparatively low rates of pay. That means city people more than suburbanites. That means lots of black people."[13]

In the 1960s, the Bell system began to hire (or promote from the operator ranks) African Americans and (in heavily Hispanic cities) Hispanic women as service representatives. In 1970, African American women composed about 14 percent of the service representative workforce in the large cities, with significant Hispanic female representation in El Paso (22 percent) and San Antonio (13 percent). Yet, few African Americans were hired directly into the service representative job. "When I applied for a job with C&P Telephone in 1970 as an African-American, I was told I qualified for an operator position. I wasn't given any information about a service representative test," Elizabeth Hargrove recalled. And two decades later, when Ameenah Salaam was hired directly into a service representative position in Delaware, "most of the Blacks in my call center were surprised. The typical career path for Blacks was to hire in as an operator, then promotion to a repair center, and then to service representative."[14]

Even as it opened up jobs to people of color, the Bell system maintained highly racialized employment practices. Black women were most heavily represented in the lower-paid operator job title. According to 1970 data, in Baltimore, Maryland, 35 percent of operators, 10 percent of service representatives, and 6 percent of technicians were Black; in Washington, DC, 45 percent of operators, 22 percent of service representatives, and 9 percent of technicians were Black; and in Richmond, Virginia, 50 percent of operators, 11 percent of service representatives, and 5 percent of technicians were Black.[15]

AT&T was (and still is) one of the largest private-sector unionized companies in the United States. In 1981, unions represented 573,000 employees at AT&T, or 67 percent of the total workforce. The largest union was the Communications Workers of America (CWA), which represented 435,000 Bell employees, followed by the International Brotherhood of Electrical Workers (IBEW), with 92,000 Bell system workers. Independent unions, organized as the Alliance of Independent Telephone Unions, represented another 45,000 Bell workers. Eventually, several of the largest affiliates of the alliance joined the CWA, including the Federation of Telephone Workers, representing 10,000 telephone operators in southern California (in 1974) and the Pennsylvania Telephone Guild (in 1985).[16]

The origin of union representation in the Bell system dates back to the 1910s, when militant female operators, concentrated in New England, organized themselves into local unions affiliated with the IBEW, along with a few local unions of linemen. In the spring of 1919, with the telephone system still under U.S. Postal Service wartime control, a wave of telephone worker strikes led by 3,000 militant New England telephone operators led U.S. Postmaster Albert S. Burleson to issue a directive granting telephone company employees the right to collective bargaining. But these efforts largely faded in the anti-union 1920s, when the Bell system established company unions through the American Bell Association that were designed to squash independent worker organization.[17]

CWA's origins date to the New Deal era. The National Labor Relations Act of 1935 banned company unions, and after the Supreme Court upheld the NLRA in 1937, the Bell system dissolved the company associations and, fearful of the AFL and CIO, in many locations actually helped or remained neutral as Bell workers transformed their company unions into independent organizations.[18] In 1938, most of the newly independent unions joined together to form the National Federation of Telephone Workers (NFTW). After a 1946 strike exposed the weakness of the NFTW's autonomous federated structure, many of the affiliated unions reorganized in 1947 into one national union, the CWA. From its founding in 1947 through 1974, CWA unsuccessfully pressed AT&T to engage in nationwide collective bargaining. Rather, regional CWA units negotiated separate contracts with the BOCs and the various AT&T subsidiaries. When CWA finally achieved the goal of national bargaining with the Bell system in 1974, CWA's founding President Joseph A. Beirne claimed this as a crowning achievement that capped almost three decades of his leadership. CWA considered the predivestiture period between 1974 and 1983, in which CWA national leaders negotiated a pattern-setting national agreement over key economic and other issues with AT&T, as a "golden age" of bargaining with the Bell system. Once the national agreement was signed, CWA regional units would negotiate the BOC contracts, and local units might negotiate what were termed "local" agreements covering issues specific to that unit.[19]

While women operators were pivotal in CWA's early years, serving as local union presidents and on the national executive board, CWA national and local leaders were mostly white men by the 1970s—even though somewhere between 52 percent and 55 percent of CWA members were women. The gender transformation in CWA's leadership ranks can be traced to the 1950s, when many female operators' local unions merged with male technician locals. In these amalgamated locals, male technicians were elected local president and dominated the executive boards while women typically served as secretary-treasurer.[20] Since local leadership was the stepping stone to regional and national elected office, by 1970, the CWA executive board composed of regional vice presidents and national officers had become an all-male (and almost all white) bastion of power.[21] And since local leadership was also the training ground for staff appointments, a 1974 CWA survey found that only 14 percent of staff was female, only four women held higher-level administrative staff positions, and only one woman served in a professional capacity at CWA headquarters.[22] Although Black people composed a significant portion of the membership, particularly among the operators, the 1974 survey found enormous racial disparity in local union leadership: only nine local presidents and sixty-five local executive board members were Black.[23]

In the 1970s, CWA women and African Americans—influenced by the civil rights and feminist movements—began to organize for greater power within the union. African American activists formed an organization, the Blacks of CWA, with elected officers and executive board members from each CWA regional district. In a 1971 letter to the CWA executive board, the organization critiqued "the almost complete lack of opportunity for advancement" for Black members and demanded that CWA adopt policies to ensure that Black people make up at least 35 percent of CWA leaders and staff, including at least one Black person on the CWA executive board. At the 1973 CWA convention, women joined with African Americans and other people of color to sponsor a resolution mandating review of a proposal for dedicated seats on the executive board for a woman and a person of color (the proposal used the term "minority"). President Beirne announced his support for the proposal in his opening convention speech, acknowledging that the all-white, all-male executive board contradicted the union's proud record of "participatory democracy." Delegate June Haskins took the microphone in support of the resolution. "I am black. I am a woman. I am a member of the Communications Workers of America. I am equally proud of all three. The rhythm of the times dictates that we should update our structure, keeping with our tradition as being a progressive Union." Fellow delegate Aleatha Pesick countered that election to union office based on sex or color, rather than qualifications, would "perpetuate discriminatory practices that the union was dedicated to eradicate." After heated debate, convention delegates turned down the resolution. Beirne closed the convention acknowledging that the union was ill prepared for the affirmative action proposal.[24]

But the pressure was on. Late that fall, the CWA executive board appointed a Female Structure Review Committee and a Blacks and Other Minorities Structure Review Committee. Many of the most militant CWA feminist leaders served on the Female Structure Review Committee, although John C. Carroll, President Beirne's assistant, was its chair. The committee met in the fall of 1973, including a "secret" caucus that excluded Carroll. As Carroll later wrote in a memo to Beirne, "This group of women had made up their minds that for once in their lifetime in the history of CWA they were not going to pay a bit of attention to any comments or viewpoints made by a male." The committee recommended structural changes to increase the number of women on CWA staff. The executive board responded that it would prioritize consideration of "carefully screened" women and minorities. The 1974 convention delegates approved a weak resolution requiring "consideration" rather than mandatory appointment of qualified women. Notably, the resolution was silent regarding Black people and other minorities, an issue that would persist for the next four decades within CWA.[25]

The 1974 convention was noteworthy for another significant advance for CWA women. With great fanfare, Dina Beaumont, president of the former independent Federation of Women Telephone Workers (FWTW), representing 10,000 telephone operators in the southern half of California, led a delegation of women, all wearing broad-brimmed hats, onto the convention floor, announcing the merger of the FWTW into CWA. The merger agreement created a new CWA district in southern California with Beaumont in the top leadership position, giving her a seat on the CWA executive board. The all-male CWA executive board now had one female member who would prove her political savvy and creativity in moving an operator—and later, a service representative—agenda within the union structures.

The agitation in the 1970s led to greater representation of women and people of color on CWA staff, but both groups continued to be significantly underrepresented in staff appointments and national elected positions. In 1982, for example, 10.4 percent of CWA staff was composed of people of color and 25 percent was female. Dina Beaumont was the only woman on the executive board. It was only in the mid-1980s, after the newly organized New Jersey state workers' units came into CWA, with its large membership of people of color, that the first African American was elected to the executive board representing CWA's public sector. Rejecting an affirmative action policy approach of strict goals and timetables, the union in 1983 created a minority leadership institute designed to train people of color to rise within the CWA structure. Yet, even as the proportion of people of color increased among the CWA membership, the national executive board did not reflect the membership. It was only after CWA adopted a constitutional amendment in 2007 creating four at-large so-called "diversity" seats on the executive board that the CWA national leadership began to reflect the gender and racial breakdown of its membership. In 2023, CWA elected Claude Cummings as its first African American president and Ameenah Salaam as its first female African American secretary-treasurer. That same year, nine women, ten African American people, and one Hispanic person were among the twenty members of the CWA executive board.[26]

EEOC Affirmative Action Consent Decree

In 1970, the Equal Employment Opportunity Commission (EEOC) launched a major assault on the Bell System's discriminatory employment practices. The landmark case, settled in 1973 with an affirmative action consent decree, set goals and timetables designed to break down the rigid gender- and race-based job segregation throughout the Bell system. While the consent decree did not eliminate gender and race-based hiring and promotion, it would significantly help open up "female" operator and service representative jobs to men, "male" technician jobs to women, and management positions to women and people of color.

Six years after Congress passed the 1964 Civil Rights Act barring workplace discrimination on the basis of race, sex, color, religion, or national origin, a full 7 percent of all EEOC complaints—1,500 in all—were filed against Bell companies, pushing the EEOC to take action. One prominent case was that of Lorena Weeks, a Southern Bell operator in Georgia with nineteen years of service who was denied a transfer request to a higher-paid "switchman" position. Weeks initially took her complaint to her local union, which refused to support her case because women were not considered the "breadwinner." The EEOC filed suit and lost in District Court, but Weeks, with assistance from the National Organization of Women (NOW), won on appeal at the Fifth District Court of Appeals in 1969. It took two more years and another court ruling for Southern Bell finally to reassign Weeks to a "switchman" position in 1971.[27]

EEOC lawyers were searching for an opportunity to take on the Bell system's discriminatory practices in one consolidated case. A young, innovative EEOC lawyer, David Copus, hit on a creative solution. He filed an EEOC challenge against AT&T's November 1970 FCC application for a long-distance telephone rate increase, alleging that AT&T was in violation of the agency's recently adopted antidiscrimination rules. While the EEOC initially targeted AT&T for race discrimination, as the lawyers delved into the Bell system's employment practices, they realized that gender was central to the companies' discriminatory policies.[28] The agency meticulously documented AT&T's gender-based discrimination in its opening brief, *A Unique Competence: A Study of Equal Employment Opportunity in the Bell System*. The EEOC case, which later was separated from the FCC rate proceeding, went on for two years, included four raucous public hearings, national mobilization by NOW, and weeks of testimony by dozens of witnesses before an FCC administrative law judge. Eventually, AT&T and the federal government signed a settlement in January 1973 that required AT&T over the next six years to open up technician jobs to women and people of color; to hire men into operator, clerical, and service representative positions; and to promote women and people of color into management positions. The consent decree included an affirmative action override that favored women and people of color over strict contractually negotiated seniority provisions for transfers and promotions.[29]

Although the EEOC repeatedly reached out to the unions to get involved, the national CWA and IBEW largely stayed on the sidelines. The union leaders and their political base came from the ranks of white male technicians who had no interest in challenging the Bell system's discriminatory employment practices. Only at the very end of the proceedings did the national unions attempt to intervene. CWA objected to the consent decree's affirmative action provisions as a violation of collectively bargained seniority rights, as well as government-mandated wage adjustments that the union asserted should be negotiated. CWA unsuccessfully tried to block the consent decree and continued to fight implementation of the upgrade and transfer plan in the years following the settlement.[30]

Although the national union opposed the consent decree, many CWA women and people of color supported the EEOC suit and provided invaluable assistance gathering evidence and delivering powerful testimony to support the EEOC case. Helen J. Roig, a Southern Bell service representative who was denied a promotion to "test deskman," described the Bell system's pervasive discrimination against women. "For female employees," she explained, "from the time you check the want ads, apply for the job, hire in, work for the Company, and retire, you are discriminated against by the telephone company because you are a member of the female sex. There will be little change for women in the telephone companies unless the companies are forced by litigation, Federal Agencies, or public opinion to change."[31]

Many CWA-represented Bell women and people of color applauded the settlement and took advantage of its provisions to move into better jobs in the Bell system. Many of the women and men who would lead the fight in later decades for good working conditions in the Bell call centers were beneficiaries of the AT&T consent decree. Some were moved into the higher-paid and previously all-male communications consultant title, a primary source of union leadership in the commercial and marketing departments. Others were among the first group of women hired as technicians who leveraged their positions among this traditional union power base to win election as union officers, next promotion to CWA staff, and then—as women union leaders—to represent the interests of the majority-female service representatives. Still others were newly hired male service representatives who emerged as union leaders and officials.[32]

World of the Service Representative before Divestiture

An AT&T ad from 1969 speaks volumes about how the Bell system saw the service representative in this period. The service representative, the ad proclaimed, "is the telephone company. At least to most of our customers, she is. . . . When someone wants a special service, an extra phone, or if there's been a billing problem . . . this is the girl they talk to. Her job is to help them. That's why she sits at a desk. Her own desk. And why she doesn't have to type or take shorthand. It's a lot easier to be helpful when you've got good working conditions. She's been to our special Service Representative school. Seven weeks' worth. So she knows her job. And she knows how to handle people." In short, the AT&T ad announced, the service representative is "our stewardess."[33]

The ad captures the contradictory nature of the service representative's position. On the one hand, she was a well-trained, intelligent professional. She had an interesting job that involved a variety of sales and service functions. She was a cut above a secretary: she did not have to type or take shorthand, she received seven weeks of specialized training, and she had her own desk. But at the same time, she was also a "girl," and judging by the picture in the ad, she was most likely a white girl. She performed her gendered role of helping people, assessing their needs and desires, calming irate customers, cajoling consumers to buy more phones or more expensive rate plans. Above all, she knew the proper methods and procedures, followed the rules, came to work every day on time, obeyed her supervisor, and thereby earned the right to be considered "special."[34]

If we were an airline, she'd be our stewardess.

She is the telephone company. At least to most of our customers, she is.

When someone wants a special service, an extra phone, or if there's been a billing problem...this is the girl they talk to.

Her job is to help them.

That's why she sits at a desk. Her own desk. And why she doesn't have to type or take shorthand. It's a lot easier to be helpful when you've got good working conditions.

She's been to our special Service Representative school. Seven weeks' worth. So she knows her job. And she knows how to handle people. Which is why we need her.

Is it any wonder we think she's special?

AT&T

If you'd like to join our Service Rep staff, call your Bell Telephone Company employment office. An equal opportunity employer.

AT&T advertisement, 1969. Source: Courtesy of AT&T Archives and History Center (Box 549–05–01). Original ad placed in *Glamour* and *Mademoiselle*, June 1969.

Most of the service representatives employed by the BOCs served residential customers. A C&P job description from 1975 outlines the major functions of the job: "handling billing and other inquiries and complaints; negotiating requests for installation, change, or removal of telephone service and preparation and/or updating of related documents, such as service orders; contacting customers, usually by telephone, to discuss billing matters; recommending and attempting to sell service and equipment to customers; and [the service representative] may prepare self-composed letters to customers."[35]

In this period, service representatives were considered "universal representatives" who handled all sales and service functions. In later years, as the number of products and services grew, the focus on sales intensified, and the geographic territory expanded, BOC and AT&T management—as well as the union—struggled over whether to functionalize the service representative job into separate job titles and channels for sales (which involved taking orders and selling products and services) and service (which involved billing inquiries and collections). Splitting the service representative job frequently included a downgrade in compensation for the non-sales position, a source of conflict with the union. Functionalization brought challenges of its own, frustrating customers with multiple transfers, making it more difficult to promote sales on service inquiries, and union disputes with management over whether management violated the contract by requiring functionalized service reps to perform duties "out of title." Universal service representatives enjoyed the variety in their job, which gave them the opportunity to exercise a wide range of skills, knowledge, and experience.[36]

During this period, the most important aspect of a service representative's job was to provide good service. "When I started working for New Jersey Bell, we were . . . under the watchful eye of the Public Utility Commission," Hazel Dellavia recalls. "We were trained to put the customer first and not sell the customer something they didn't need, because they would remember it and have a negative opinion of the company. We were observed for our ability to satisfy the customer, our tone of voice, and production." Ronald Collins, who began working as a service representative for C&P Telephone in Baltimore, Maryland, in 1981 after the 1973 affirmative action consent decree required the Bell system to hire men into this formerly all-female position, has similar recollections. "They really focused on customer service . . . treat the customer well . . . take as much time as you need to make that customer happy." Elinor Langer, who wrote a piece in the *New York Review of Books* in 1970 about her three-month stint as a New York Telephone service representative in training, recounts that her training instructor taught the "Customer Service Ideology" using this example: "If the customer tells you to drop dead, you say 'I'll be very glad to help you sir.'"[37]

State regulatory commissions set the rates that BOCs could charge customers and monitored the company's quality of service. The commissions did not look kindly on a request for a rate increase if there was evidence that the telephone company scrimped on service. This deal between regulators and Bell company management—a guaranteed return on investment in exchange for universal, affordable, and quality service—fostered a deep culture of service throughout the Bell system. The companies' voluminous, detailed methods and procedures manuals, extensive training, and close supervision were designed to standardize service representatives' interactions with customers to meet high standards of service. While customer service representatives deeply resented policies that they believed treated them as children rather than adult professionals, they also felt that management's customer-focused work culture aligned with their own commitments and interests to provide good service to customers.

The business offices were small, typically twenty-five employees, and handled customers who lived in the local area. At that time, according to a former business office manager in upstate New York, "the people who serviced you were your neighbors, your friends, everybody knew everybody." This would change in future years, as Bell system executives consolidated business offices into larger call centers, distancing the service representatives from the communities they served.[38]

Service representatives sat at desks positioned to face each other. Each had a rotary phone to receive and place calls, a large paper manual called the Service Rep Handbook, and an outbox for service orders and other paperwork ready for distribution to other departments. The telephone was connected to a jack with a headset that the service representatives wore. Customer records were all paper. They were kept in large filing cabinets called "tubs" located next to the service representative's desk.[39] The picture on the next page shows a male African American service representative searching for a customer record in a tub next to his desk. The depiction of an African American male service representative is misleading. There were few male African American service representatives at the time. Illinois Bell provided the picture to the EEOC to show that at least some African American men had this job title.[40]

Customers dialed a local number to reach the business office serving their area. A telephone operator would take the call and route it to the service representative responsible for the customer's exchange, or, if that representative was busy, to another representative in the office. The

Illinois Bell customer service representative, circa 1970. Illinois Bell submitted this photo as evidence in the Equal Employment Opportunity Commission discrimination suit against AT&T. Despite the photo, there were few male African American service representatives at that time. National Archives at College Park, MD. Docket #18143, Docketed Case Files 1927–90, RG 173, Box 163:36 (AT&T/Equal Employment Opportunity Commission case).

service representative would locate the customer's record in the appropriate tub. Frequently, this meant the service representative had to get up and walk over to another representative's tub to get the record. "I'd put the customer on hold, physically get up, go over, get the customer record. It was nice to get up and move around a bit," Collins recalls. "We walked around constantly. It was much more social," recounts Dellavia. "Getting up to retrieve records from other tubs allowed us to chat with each other throughout the day," former New Jersey Bell service representative Linda Kramer explains. But, she adds, this was a practice that the supervisors tried to curtail.[41]

The service representative would write up orders by hand, with a separate code for each function, and put the handwritten order in the outbox for pickup by a clerk who brought the order to the service order typist. A clerk would later return the typed order to be filed by the service representative in her tub. On a typical day, the service representative handled thirty to sixty calls. The office was open from 9 a.m. to 5 p.m., Monday through Friday. The representative worked seven or seven and a half hours a day with two fifteen-minute paid breaks and an unpaid lunch hour. Unlike operators, the service representatives did not work split shifts, weekends, or holidays and had a predictable nine-to-five schedule. Each representative would receive a daily schedule, which would include anywhere from thirty minutes to two hours off the phones to make outbound calls to customers delinquent on their bills (called "treatment" calls) or to handle the paperwork for complex orders. Elimination of this "closed key" time (as it was called) became a central strike issue at Verizon in the year 2000.[42]

Every step of the service representative's job was prescribed by detailed Bell system methods and procedures. A 1965 Bell of Pennsylvania Analysis of Job Requirements spells out exactly what the service representative was required to do. On a "customer initiated contact," she must prepare a "Contact Memorandum" form 3882, and if she must leave her desk to get an answer to a question, she must return within 90 seconds. On service orders, she must prepare form 3882-1. On new and transfer orders, she must "make four attempts to sell monthly rate items" or "make it up on another contact." On new connections, she must prepare Customer Credit Record form 3745-2 to be sent to the accounting department, send the stub to the customer, and give the customer a commitment date for service. When a customer denies all knowledge of a call on a bill, she must prepare an Adjustment Memo form 3730-2, apologize to the customer, tell them to deduct it from the bill, and send the memo to another department for investigation.[43]

While the focus of the job was on service, Bell managers also expected service reps to "upsell" to customers. To be sure, there were not many products and services to sell, but even so, service representatives were given a sales quota for princess or trimline phones and packages that included multiple extensions in a home. At New York Telephone in 1969, there were monthly sales contests with $25 cash prizes, coffee and donut rewards for high sellers, or more chances to win a raffle for a free turkey at Thanksgiving.[44] While representatives were encouraged to sell, they were not disciplined for failure to meet sales quotas as long as they made the effort to sell. This would change drastically in the post-divestiture period.

Training for the BOC service representative job usually involved eight to ten weeks before assignment. Supervisors gave further training, both in the classroom and on the job, after assignment. "Training during the Bell system era was extensive and thorough," Dellavia remembered. "Supervisors were familiar with the work and were able to give you tips on how to deal with unusual situations." At that time, most frontline supervisors were promoted from the service representatives' ranks.[45]

The BOCs were responsible for customer care for AT&T Long Lines' residential customers. It was not until the post-divestiture period that AT&T set up its own residential consumer customer service operation. However, AT&T Long Lines did have a customer care division to service business customers who purchased private line network services. (Private lines were dedicated circuits connecting multiple locations belonging to a single company or large institution, creating a virtual private network.) About 1,500 AT&T service representatives worked in several dozen offices around the country, with about five to twenty-five representatives in any one office.[46]

Before 1980, the AT&T service representatives occupied a strange position as "advisory employees" who were represented by but not eligible for membership in the union. The situation dates back to 1948, when CWA and AT&T agreed to resolve a dispute over bargaining unit representation by giving these job titles union representation but not eligibility for union membership. As the customer service workforce at AT&T Long Lines increased, CWA local unions pressed the national union to change this anomalous situation, and in 1980, the company agreed. Within three years, the local unions that represented these service representatives were demanding higher pay for the service representatives, explaining that "when we finally corrected the error (barring service representatives from union membership) . . . we opened the door for these employees to demand proper compensation for the work they performed."[47]

The service representatives found great satisfaction in solving customers' problems and meeting their needs, and they particularly enjoyed the variety of work and the comradery of the office. "I liked helping people," Hazel Dellavia recounted. "My satisfaction came from doing the best I could to help the customer. Even when they were nasty, I tried to be sympathetic and calm them down so I could help them." Gail Evans liked the variety. "It was like reading a book. The next customer comes in and they tell you everything. One was a billing call, the next one was sales, the next one is totally different . . . a different chapter with every call." Linda Kramer liked the fact that the job was close to home, but even more, she cherished the lifelong friendships she made with her coworkers. "We did a lot together outside the office. Many are still my friends." Sandy Kmetyk appreciated the fact that the company "considered us important. We were special. We took pride in our jobs. The operators considered us prima donnas." Barbara Fox Shiller enjoyed the comradery. "We built close relationships. You would sit next to somebody and you could talk. We could talk about work. We could vent. We could share information. And we could get up and walk around to get records from different tubs."[48]

But service representatives also resented the detailed rules and procedures. "Having to get permission to use the bathroom was a particular peeve of mine," Dellavia recalled. "I understood it when I was an operator. You couldn't just walk away and leave the calls hanging. But in the business office, you could turn off your phone and no calls would come in. Instead of giving the reps the dignity of working this out among themselves, you had to get permission from a supervisor. In a room where everyone could see everyone else and you had a light at your position that went on when you were open for calls, I thought this was really stupid."[49]

Supervisors listened in on a sampling of the service representatives' calls. "Sometimes they would listen in a tone room, which was a location in the building, but you never knew whether a supervisor was in there," Ron Collins explained. "Sometimes they would listen from their desk. Sometimes they would listen sitting next to you. The supervisor would get with you soon, usually within a day or so, and they would give you feedback that this was great, you did everything you were supposed to, you may have been able to do this better." The service representatives were rated on such factors as how long it took to answer the call, how well they greeted the customer, responsiveness to customer need, knowledge of the services and products, customer satisfaction, tone and manner, and sales offers. Supervisors were required to observe a minimum of five calls per month for each representative. While the managers put the results of the observation in a service representative's personnel file, and while the evaluation could affect promotion opportunities, a service representative was not disciplined based on these observations. The only exception was if the service representative was "abusive" to a customer. This would change dramatically in later years.[50]

Bell system managers placed top priority on attendance and timeliness. Typically, more than six or eight absences or ten instances of tardiness in a year were considered grounds for dismissal. Attendance records indicate that most Bell employees, including service representatives, came to work every day and that attendance was not a significant problem. In 1972, the absence rate for Bell system employees in the departments where service representatives worked was just 2.9 percent per year, only slightly higher (by just one percentage point) than those of their male colleagues. The low absence rate would change significantly in the post-divestiture period with the increase in stress-related illness among the service representatives.[51]

Most service representatives were young, and among this group, turnover was high. In 1968, 12,000 of the 28,000 service representatives in

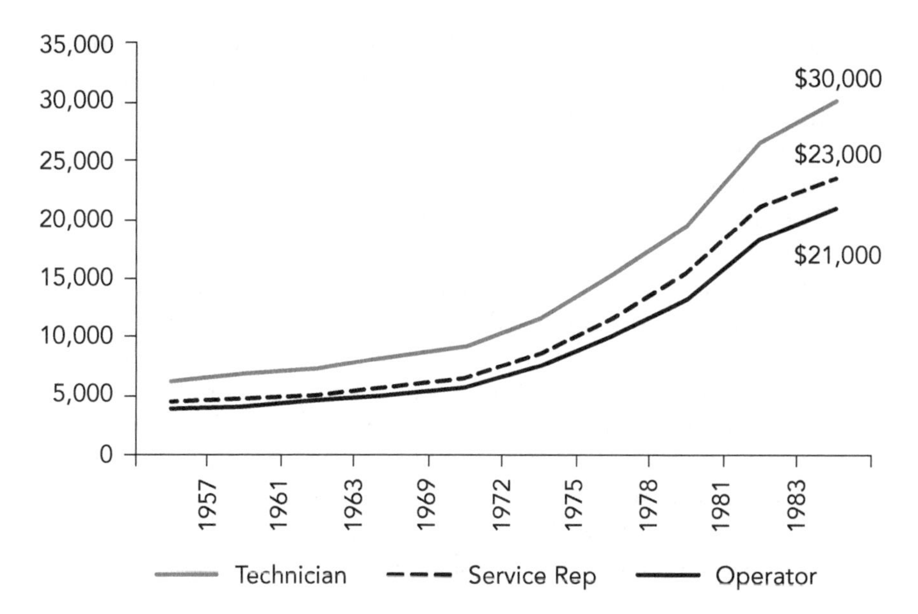

Bell Telephone System Annual Wages by Occupation, 1957–1983. Sources: CWA/C&P Telephone contracts; John Strouse; 1974 Bell Systems Bargaining book.

the Bell system, or 42 percent of them, had been hired and trained that year. The high turnover was concentrated among those with less than six months' job tenure. AT&T estimated the economic impact of just 1 percent turnover among service representatives at $1 million a year.[52] Despite the high turnover among newly hired service representatives, a significant number of representatives stayed on the job for many years, perhaps taking some time off to raise young children or follow husbands to another city, only to return to work after a few years when the children were older, family circumstances changed, or marriage ended in divorce. In 1971, for example, average tenure in the commercial departments was 8.9 years, and more than one-third (31 percent) of employees had more than ten years of service.[53]

The unionized service representatives earned good wages compared to other U.S. working women, and certainly compared to other U.S. working women with comparable education levels. In 1983, a Bell service representative with four or more years' experience earned about $24,000, which was twice the median annual earnings for full-time women workers with a high school diploma ($12,841) and almost twice that of all full-time women workers ($13,915). Economists have identified structural pay disparities in most female-dominated occupations in the United States. But in what a leading feminist economist would later call an "exception to the rule of low-pay for women's work," the 1983 union-negotiated wage rates for Bell system service representatives ($24,000) was higher than those for male high-school graduates ($20,536). Collective bargaining provided the Bell system service representatives a mechanism to overcome gender-based pay discrimination in the broader economy.[54]

While collective bargaining provided Bell women in the female-dominated occupations better wages than their sisters in the larger economy, it did not eliminate gender-based pay discrimination within the Bell system itself. Negotiated salary schedules privileged the male-dominated technician jobs throughout the Bell system, reflecting a devaluation of the complex customer contact skills, emotional labor, and deep knowledge required of a customer service representative, as well as the market-based pay inequities that women workers faced throughout the economy. Year after year, Bell system service representative wages trailed technicians' earnings by between 20 percent and 30 percent.[55]

The Uprising of the Operators

The telephone company operators were the first to experience the negative impact of digital technologies on working conditions. The Bells began to deploy electronic switching systems, computerized operator terminals, and associated software in the operator services departments in the 1960s, several decades before these systems migrated to the customer service offices. The electronic systems degraded the directory assistance operators' job to pushing a few buttons to transfer callers to automated response systems. The new technology also gave management the ability to embed machine pacing and electronic surveillance in the algorithms that simultaneously routed the telephone call, determined the operators' speed of answer, and surveilled her work performance. Operators experienced the resulting loss of control compounded by higher demand as an increase in stress at work.

The Bell system had always closely monitored operators' work performance for quality and cost-saving efficiencies. Since the 1890s, supervisors (often on roller skates to ease movement down the line) stationed themselves behind the operators to listen to their conversations with customers. Later, when the telephone companies introduced electromechanical switching in the 1920s, ushering in the direct dial era, engineers wired the switches to capture detailed information about operators' call handling and answer time. Managers used this information to reduce operator downtime, resulting in a 20 percent increase in operators' workload.[56]

The intensification of surveillance, speed-up, and deskilling that accompanied the introduction of digital systems prompted the operators to demand that their union take action to curb these highly stressful practices. The result was an uprising of the women within CWA, as they married their struggle for greater power within the union to their campaign to alleviate the conditions creating intense pressures on the job. Although the customer service representatives were ancillary players in these struggles, due to their smaller numbers and less stressful working conditions, the call center workers would later build on the foundation that the operators created as a result of their thirteen-year campaign to curtail the detested surveillance of their work.[57]

The CWA-represented female operators—just as their service representative sisters in later years—faced a double challenge. First, they had to organize within their male-dominated union to put their concerns about job pressures on the CWA agenda. The work environment of the male technicians stood in stark contrast to the regimented operator workplaces. The technicians experienced a large degree of autonomy and control at work, fostering independence and a militant work culture. Not only were they physically separated from direct supervision as they moved from job to job, but they also possessed specialized craft knowledge.[58] The predominantly male CWA leadership was ignorant, at best, and unconcerned, at worst, about the job pressures the female telephone operators experienced. Therefore, CWA operators and the women leaders who represented them first had to launch a campaign *within* the union to put their issue of job pressures on the agenda and make them union priorities.

Second, the female operators and union leaders had to craft solutions that challenged "management rights" to organize the business and the workforce. Traditional union mechanisms proved inadequate to protect operators against technology-enabled work speed-up and disciplinary action. There was no contract language restricting management's right to implement new technology, to restructure work and job titles, to monitor employees' conversations or workloads, or to use the information gathered to discipline employees. With no contract language to restrict such practices, the union could not turn to the grievance procedure to protect members. Before the job pressures campaign was launched, the union had never raised the issue of monitoring, time indexes, or other workplace job pressures at the bargaining table.[59]

CWA women leaders launched their job pressures campaign in 1967 at the twenty-eighth annual CWA convention. The delegates unanimously approved a resolution mandating that the CWA executive board "conduct a study of the undue pressures being applied throughout the Bell System to traffic operating employees and to appropriate measures to alleviate such pressures."[60] ("Traffic operating employees" refers to telephone operators.) After the convention, the CWA executive board created a job pressures study committee. A committee survey of CWA members and leaders found that operators overwhelmingly reported highly stressful conditions because of productivity measures called "index systems," job observation, and rigid absence control policies, while male craft workers were largely unaffected by these forms of supervisory control. Although the vast majority of survey respondents were operators or technicians, a small percentage were service representatives. Some shared the operators' frustration. "[Our managers] have trained us all as service representatives for eight weeks to think for ourselves and make our own decisions," one service representative wrote. "Why not let us exercise our judgment?"[61]

The job pressures committee reported its findings to the CWA executive board and shared a summary with delegates to the thirty-first CWA convention. The convention took no action. Subsequently, the CWA executive board appointed a three-person job pressures implementation committee; the committee urged President Beirne to make alleviation of job pressures, in general, and secret punitive monitoring, in particular, union priorities with a nationwide member mobilization. The executive board failed to act on the proposal, disbanded the committee, and let the issue wither until newly energized and militant CWA women put job pressure issues squarely on the union agenda at the first CWA national women's conference in 1978.[62]

Dina Beaumont, the only woman on the CWA executive board, chaired the CWA Concerned Women's Advancement Committee, organizers of the four-day event designed to teach CWA women skills to "move up the ranks of the Union." The conference, attended by hundreds of CWA women, was structured as a mock CWA convention to teach participants parliamentary procedure, public speaking, committee building, and resolution writing. Conference participants approved a job pressures resolution, which called for a CWA national day of mobilization against job pressures. The CWA executive board subsequently scheduled a national Job Pressures Day for June 15, 1979.[63]

On that day, CWA locals rallied across the country under the slogan "we are people, not machines." They used humor and fun to attract attention. In Fort Collins, Colorado, employees and members of the public were invited to "soak away their job pressures" in a hot tub. In Cleveland, Ohio, workers took a "whack at Ohio Bell" with a sledgehammer and an old Ohio Bell car. In Oakland, California, a jazz band played "When the Saints Come Marching In" as the local "buried" an operator who died from job

Assembly lines are only in factories. Right? *Wrong!*

The employees in the picture on the left are as much a part of an assembly line operation as the employees in the picture on the right.

In both cases, they are constantly pushed toward greater and greater productivity, are subjected to over-supervision, harsh absentee control programs and they must adhere to strict, unyielding schedules.

The telephone worker in your town and the auto worker in Detroit have a lot in common: Both are an appendage to a machine.

Your telephone operator—like millions of other American workers—suffers from job pressures . . . pressures that exact a toll both on the worker and his or her family and friends.

Many telephone operators have a pleasant work environment—bright lights, soft music, a comfortable chair and a carpet on the floor.

But, so long as there is monotony, boredom and repetition — machine-like qualities—job pressures exist . . . whether you work in a phone factory or a car factory.

We want the bosses of business to know:

"We are people, not machines."

CWA Job Pressures Day flyer, 1979. Communications Workers of America Records, WAG 124, box 159, folder 1. Tamiment Library/Robert F. Wagner Labor Archives, NYU Special Collections, New York University.

pressures. Georgia Governor George Busbee, Atlanta Mayor Maynard Jackson, and the Los Angeles City Council issued proclamations in support of CWA's assault on job pressures. News media reported on the events. Local unions distributed a nationally produced flyer to educate the public that telephone employees were "as much a part of an assembly line" as autoworkers. "They are constantly pushed toward greater and greater productivity, subjected to over-supervision, harsh absentee control programs and they must adhere to strict, unyielding schedules," the leaflet announced. Although telephone operators might have a pleasant work environment, "machine-like qualities" of monotony, boredom, and repetition created unbearable job pressures.[64]

National Job Pressures Day, along with a CWA technology conference that same year, finally succeeded in putting the issue of supervisory monitoring on the national union's bargaining agenda. In 1980, the CWA Bell system bargaining council adopted, for the first time, the demand to "eliminate monitoring" as part of that year's collective bargaining priorities. The 1980 bargaining minutes contain no reference to discussions over this demand. Yet, when the negotiations concluded, CWA achieved what it termed a "major victory" as AT&T agreed to ban secret diagnostic monitoring of operators, and to use observations for training—not disciplinary—purposes. In a letter from AT&T labor relations chief Rex R. Reed to CWA President Glenn Watts, AT&T agreed "as an indication of our mutual determination to achieve goals of improvement" that all diagnostic performance monitoring would be performed "at the position where the individual being observed is working when management determines it is necessary for training or instructional purposes." This language was incorporated into the Bell Operating Company contracts. While on its surface, the language reads as weak, in reality, by requiring that operators know when they were being observed, the agreement effectively ended the use of monitoring to "catch" an employee, thereby significantly reducing stress. The Reed-Watts letter on supervisory monitoring represented what CWA national leadership proclaimed as a major victory for telephone operators. The contract language did not cover the customer service representatives, who would have to mobilize for such protections in later years, as managers introduced management practices from the operators' workplaces into the service representatives' call centers.[65]

Bargaining for Codetermination

The Reed-Watts monitoring letter was one piece of a much larger package that CWA negotiated in the 1980 agreements with AT&T and the Bell

Operating Companies, provisions that opened the door for greater union voice over the introduction of new technology and the organization of work. Technology change was a constant in the telecommunications industry, and CWA leaders had long recognized that the union's role was to negotiate provisions that enabled their members to retrain for the new jobs and to garner economic benefits from rising productivity.[66] In 1979, a CWA research department paper predicted that the introduction of digital switches, automated network monitoring and provisioning systems, and increased competition would lead to massive job dislocation for technicians, operators, and clerks, and significant growth in the number of sales and service workers in the Bell system. The paper emphasized that automated systems could deskill some jobs while adding increased responsibility and skill requirements to other positions. CWA leaders and national staff studied German and Scandinavian labor relations systems and sociotechnical theories of workplace technology and work organization, and they determined that in 1980 bargaining, the union would press for greater union involvement in business decisions regarding implementation of technology and work organization to protect members' job security and reduce job pressures.[67]

The result was what CWA leaders called a historic agreement, first with AT&T and extended to the BOC contracts, which, along with the monitoring protections, established joint labor-management committees on technological change, quality of work life, and job evaluation. These provisions were designed to address the "interrelated issues of job security and job pressures" triggered by technological change and an ailing economy. The technology change committee was designed to "eliminate the detrimental effects of changing technology on the workforce" by mandating at least six months' advance notice to the union of major technological change, and providing a forum to discuss training, employment, and early retirement opportunities. The joint working conditions and service quality improvement committee aimed to give "workers an opportunity to participate in the design and implementation of their own jobs," leading to the adoption of a quality of work life program, discussed in greater detail in chapter 4. Finally, a joint occupational job evaluation committee was established to develop procedures for reclassifying jobs or instituting new job titles.[68]

CWA saw great potential in these provisions. In the words of one CWA leader, these agreements contained the seeds of European-style codetermination, giving the union contractual rights to participate in key business decisions. While the agreements were extended in 1983 (and remain in the contracts today), the dislocations caused by the 1984 AT&T divestiture disrupted what might have evolved into a more mature collective bargaining relationship between CWA, AT&T, and the divested BOCs. In 1983, 500,000 CWA members went on strike for fifteen days when the union and Bell negotiators were unable to reach agreement over a contract that would frame the wages, benefits, and working conditions for the post-divestiture era. The progressive contract provisions giving the union a greater voice in new technology and workplace operations would face challenges in the post-divestiture era, as AT&T and CWA struggled over the company's strategic approach to business prosperity in a more competitive era. The organization of work in the strategically important customer service call centers became a flashpoint in this struggle.[69]

In the pre-divestiture Bell system, customer service representatives worked in a stable business environment, selling and servicing one product—voice telephony—in small offices, keeping paper records, getting up multiple times a day from their desks to move their bodies and chat briefly with coworkers. They found satisfaction using their skills, knowledge, and emotional intelligence within the Bell culture of high-quality customer service. To be sure, the Bell system's detailed bureaucratic rules and supervision forced conformance, but absent the technical control of automated workforce management systems, they had a measure of autonomy over the pace and manner of work.

But a revolution was coming, driven by the competitive forces unleashed by the breakup of the Bell system and introduction of digital technologies with the capacity to intensify the speed, surveillance, and even scripting of the service representatives' work. It would begin in the brand-new AT&T call centers and spread to the Bell Atlantic business offices. The customer service representatives organized within their union to contest the new terrain, taking lessons from and building on the operators' campaigns to relieve job pressures. Their fights for dignity, control, healthy working conditions, and wages that were commensurate with the value they created for their employers would take many forms—collective bargaining, organizing, labor-management joint programs, and even a strike for stress relief—as their employers transformed their companies from public service monopolies to profit-maximizing financial organizations. And in the process, they transformed themselves into a workforce of resistance.

CHAPTER 2

Becoming a Workforce of Resistance

Mary Ellen Mazzeo, in a freshly ironed blouse and dress pants, walked into the brand-new AT&T call center in Syracuse, New York, filled with excitement and trepidation. It was 1984, the year of the AT&T divestiture. The company was creating an entirely new customer service operation, and she wanted in. Bored after twelve years as a New York Telephone Company operator, she was looking for more challenging work with higher pay. Mazzeo accepted a job as a sales representative at the new AT&T long-distance company. "The job was stressful in the beginning, learning all the acronyms, not knowing what I was doing, trying to do a good job," Mazzeo recalled. In her first years, the job pressures came from the complexity of the work and Mazzeo's drive to give good service. But by the early 1990s, as AT&T increased sales quotas, negotiated commission pay plans, and intensified electronic monitoring, the working conditions drove high levels of stress. "Many of my coworkers would dread Sunday nights before going back to work," Mazzeo remembered. "They would fight so hard to meet quota by the end of the month. They were put on performance improvement plans. Some were fired. It became more and more difficult to do the job." Mazzeo became active in the union, and as the call center grew to more than 500 sales representatives, she rose to become president of Communications Workers of America (CWA) local 1152 and a national leader representing the union's customer service members in their fight for good jobs at AT&T.[1]

Unlike Mazzeo, who moved to AT&T at divestiture, Judy Buchanan, with eighteen years as a service representative at C&P Telephone, stayed with the local company. But her job at the Bell Atlantic Regional Bell Operating Company (RBOC) subsidiary did not remain the same. Rather, as she wrote in a letter to the 1986 CWA bargaining committee: "Never in my 18 years have I been subjected to the pressure, tension, and stress in my job that I've been facing in the last 2 years. The company has but one thought in mind and that is to ball and chain the once flexible service rep to the black box on her desk. They are now putting screens around our desk to isolate us from each other so our FULL CONCENTRATION is centered entirely on SELL-SELL-SELL. Our job has grown by leaps and bounds and the stress and tension have grown in proportion. Absenteeism is up and nerve related problems plaque [*sic*] our workers. Morale is at an all-time low."[2]

Mazzeo and Buchanan were two of the tens of thousands of customer service representatives who experienced the transformation of the labor process and downward pressure on living standards in the decade after the Bell system breakup. Two major developments came together to revolutionize their working environment. First, neoliberal policy that favored competition over monopoly regulation led to the 1984 AT&T divestiture. The radical restructuring of the telecommunications industry unleashed intensive competitive pressures, particularly in AT&T's long-distance market, driving management to slash its most variable expense—labor—and dramatically increase sales pressures. Second, technological advances that integrated computing with digital communications networks created new products and services. These new revenue opportunities, along with the more competitive market structures, elevated the importance of the customer service employees while simultaneously giving managers new tools to surveil, deskill, and control the labor process. Managers used the machine-paced algorithms built into the automatic call distribution equipment to measure the service representatives' call servicing time, eliminate downtime between calls, and thereby speed up the work. The customer service workplaces began to look more like the highly regimented, automated operator service centers that the Bell system had long organized along a Taylorist scientific management model. Just as operators mobilized a decade earlier against the deskilling, speed-up, surveillance, and job pressures associated with the introduction of digital technology, the service representatives began to organize within their union to resist the stressful degradation of working conditions in their call centers. In the process, they coalesced into a workforce of resistance.[3]

In this chapter I focus on collective bargaining strategies that the union and its customer service members deployed in their contest with their call center managers. Collective bargaining is a power struggle between the union and the employer, shaped by the members' level of organization and mobilization, as well as the economic context within which the employer operates. Contract negotiations, then, become an arena for worker resistance as the union contests with the employer over the distribution of the profits (value) that employees create, over the conditions at work, and over the pace and manner in which the work is done. When the Bell service representatives mobilized as an occupational group to demand that their union prioritize their issues at the bargaining table, they became a workforce of resistance. It is this process and the demands they made for contractual protections that I describe in this chapter.

As we shall see, the service representatives at Bell Atlantic proved more successful than those at AT&T in their struggles for good working conditions and compensation. Several factors explain the different outcomes. First, the two companies operated in very different market environments. During this period, the Federal Communications Commission (FCC) moved aggressively to promote competition in AT&T's long-distance markets, while state commissions continued to regulate the rates and service of Bell Atlantic's local monopoly. As a result, AT&T swam in treacherous waters, losing market share every year to MCI and Sprint, whereas Bell Atlantic operated in much more protected seas, particularly in the consumer segment. Second, AT&T adopted the newest and most powerful technology in its brand-new call centers, whereas Bell Atlantic more gradually transitioned to fully automated systems. Finally, CWA had to discover the locations of the new, unrepresented AT&T call centers, building the union in these centers from scratch, whereas the Bell Atlantic service representatives inherited the local union structures and bargaining relationships from the pre-divestiture era.[4]

AT&T Divestiture

On January 8, 1982, AT&T CEO Charles L. Brown and U.S. Department of Justice (DOJ) Assistant Attorney General for Antitrust William Baxter signed the historic agreement to break up AT&T in what one analyst called the "biggest, most complex restructuring in the history of business."[5] The divestiture, which affected $100 billion in assets and almost 1 million employees, took effect on January 1, 1984. The legal agreement, known as the *Modification of Final Judgment*, required AT&T to spin off its local telephone network, employees, and assets, creating seven independent Regional Bell Operating Companies (RBOCs), each with about $20 billion in assets. The local companies were barred from long-distance service or manufacture of telecommunications equipment. The new AT&T was restricted to long-distance, telecommunications equipment manufacturing and installation through its Western Electric subsidiary, and research and development at Bell Laboratories. The consent decree offered AT&T two carrots: first, it terminated a 1956 ban on AT&T's entry into the computer business—the future of communications, in AT&T Chairman Brown's view—and second, it relieved AT&T of responsibility for major capital investments to upgrade local analog networks for the digital age. Most significant, as business historians Peter Temin and Louis Galambos wrote, "Divestiture dethroned the national integrated network in favor of competition."[6]

While new technologies created the conditions for regulatory reform, and intensive lobbying by big business pressed the FCC to foster competitive entry into the long-distance and equipment markets, it was policymakers' faith in neoliberal ideology that grounded their decision to abandon New Deal–era public oversight of infrastructure industries in favor of deregulation and free market competition. Residential and small-business customers were not clamoring for change, while big business focused on competition policy, not breaking up the Bell system. Certainly, the emergence of new microwave and switching technologies opened the door to competitive entry in the long-distance and equipment markets, which the FCC sanctioned first with the 1968 *Carterfone* decision allowing connection of non-Bell equipment to the network and the 1969 and subsequent MCI rulings permitting competitive entry into long-distance telephone service using microwave technology (MCI stands for Microwave Communications, Inc.). Yet, as Temin and Galambos argue, it was "ideology, not technology, that triumphed in the 1970s and 1980s."[7]

Since the 1950s, University of Chicago economists George Stigler, Ronald Coase, and Milton Friedman had been laying the intellectual foundation for deregulation, but it took the economic crisis of the late 1960s and 1970s to create the political will to deregulate railroads, airlines, trucking, and, most relevant to this study, the breakup of the monopoly telephone company. In searching for solutions to the economic problems of the period, characterized by slower economic and productivity growth, rising inflation, high unemployment, and stagnant wages ("stagflation"), policymakers looked for fresh approaches to replace New Deal programs of regulated capitalism.[8] Many were persuaded by economist Alfred Kahn and future Supreme Court Justice Stephen Breyer, among others, that price and entry regulation of infrastructure industries led to higher prices, inefficiencies, less investment, slower growth, and regulatory capture by politically powerful companies and unions.[9] With Kahn at the helm of the Civil Aeronautics Board and Breyer as lead counsel of Senator Edward Kennedy's investigation into airline regulation, Congress passed the Airline Deregulation Act of 1978, with deregulation of trucking and railroads two years later.[10] Support for deregulation of infrastructure industries was a bipartisan affair. Democratic majorities in Congress pushed for and Democratic President Jimmy Carter signed the legislation that overturned government oversight of price and entry regulation in transportation industries, disrupting stable market structures, employment, and labor-management relations. And though AT&T divestiture took place under the Republican administration of President Ronald Reagan, Democrats as far back as the Johnson administration and Democratic leaders in Congress had pushed for regulatory reform of the telecommunications marketplace.[11]

Telecommunications workers were the sacrificial lambs of the divestiture decree. The former Bell companies cut about 200,000 union jobs in the decade after the break-up. The largest job cuts were at AT&T. Because the union contract made it difficult to reduce wages and benefits at AT&T, the company cut labor costs by eliminating more than 125,000 technician, operator, and Western Electric factory worker and installer positions. Many employees who remained with the company were forced to move multiple times as AT&T consolidated work locations. At Bell Atlantic, the company cut about 10,000 telephone jobs, mostly through attrition in the first few years after divestiture, split evenly between management and union positions. (The other RBOCs together eliminated another 65,000 union jobs.)[12]

While many Bell customer service employees remained with their former employer, others considered seeking a transfer, making educated guesses as to which company was more likely to prosper and provide job stability or advancement opportunity. For those who remained with an RBOC, the gamble paid off with relatively more job security, whereas many who worked at AT&T faced difficult choices as the company consolidated and reorganized its operations multiple times.[13]

The AT&T breakup disrupted the relatively stable labor–management relations of the monopoly era, leading to strikes against AT&T in 1983 and 1986. Union density at AT&T plummeted from 62 percent to 25 percent from 1984 to 1995 as AT&T cut union jobs, moved many technical jobs out of the bargaining unit, and walled off acquisitions and new subsidiaries from union representation.[14] In contrast, greater stability in the local telephone market at Bell Atlantic and other RBOCs led to little change in union density over this period; it remained high, at about 65 percent to 70 percent.[15] (Later, as Bell Atlantic grew its wireless business and merged with other entities, the company deployed the same union avoidance tactics and experienced greater union unrest and marginalization.)[16]

The irony of telecommunications reform was the enormous amount of FCC regulation required to hobble AT&T and inject competition into the

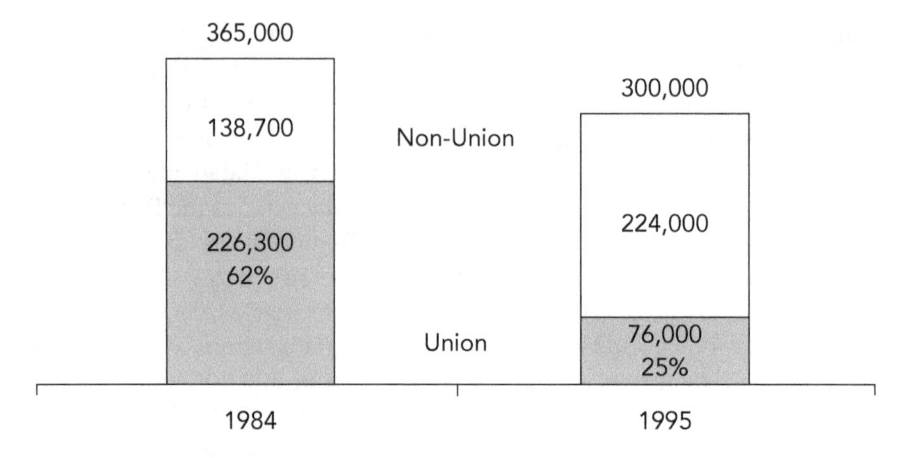

AT&T Union and Non-Union Employment, 1984 and 1995. Source: AT&T SEC Form 10-K

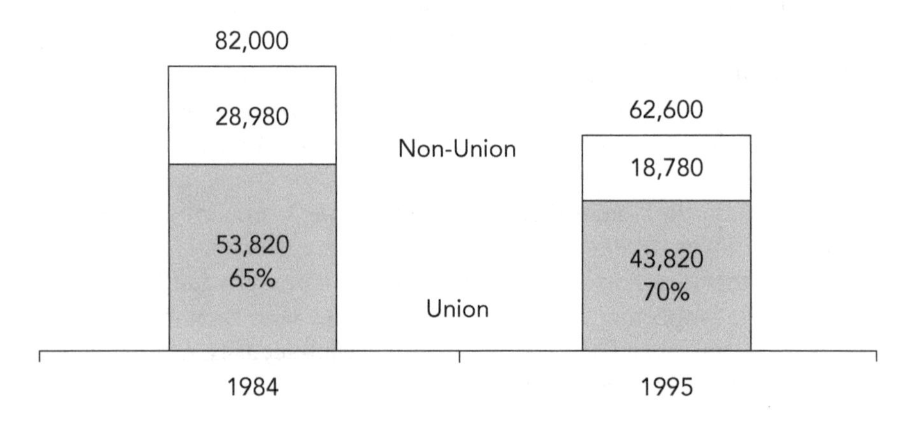

Bell Atlantic Union and Non-Union Employment, 1984 and 1995. Source: Bell Atlantic SEC Form 10-K

long-distance sector.[17] After divestiture, the FCC required long-distance companies to pay access charges to the local telephone companies to use the local network, and to replace monopoly-era subsidies. The FCC gave MCI, Sprint, and other new entrant long-distance companies a 55-percent discount from the AT&T rate, providing the new companies a huge cost advantage over AT&T. The FCC continued to regulate AT&T's rates through the mid-1990s, whereas MCI and Sprint were not subject to rate regulation. Moreover, the FCC mandated a two-year period known as "equal access," which required outreach to every customer to select a long-distance carrier.[18]

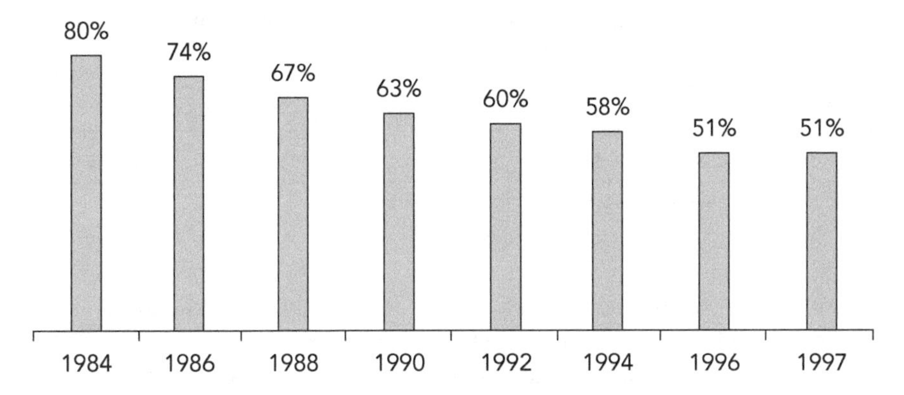

AT&T Long-Distance Market Share, 1984–1997. Source: FCC, Statistics of Common Carriers, Table 8.18, 1996/97

State public service commissions continued to regulate Bell Atlantic's telephone rates to protect consumers in what was considered a monopoly market for local telephony. At the same time, the regulators allowed the company to set the price of other services, including inside wire maintenance plans, custom features such as call waiting and caller ID, and other specialized services. In the mid-1990s, state regulators began to transition to a new regulatory regime that capped prices rather than return on investment, rewarding the telephone company for cost-cutting measures. State commissions introduced competition into the business sector by granting new entrants licenses to build networks in urban centers or for large business locations, cutting into Bell Atlantic's lucrative business market.

Most analysts expected AT&T to prosper and the RBOCs to stagnate after divestiture, but the opposite took place. The new AT&T struggled to find a winning competitive strategy. Given the competitors' regulatory cost advantage, coupled with their state-of-the-art new fiber networks and much more lean management structures, AT&T lost about 40 percent share in the consumer and small-business long-distance markets in the decade after divestiture. After more than a decade of competitive entry into the lucrative equipment market, AT&T's share stood at only 12 percent in 1987.[19]

While AT&T remained a corporate giant, with annual revenues topping $60 billion to $70 billion a year, the company was never able to realize Chairman Brown's dream to become a leading computer company. In 1966, then AT&T CEO Bob Allen conceded defeat of the company's computer strategy by selling its NCR computer business (acquired only four years earlier) at a loss of $1.6 billion.[20] In contrast, Bell Atlantic's regulated

Table 2. AT&T and Bell Atlantic Financial Performance, 1984–1995

	AT&T	Bell Atlantic
Revenue Growth	32 percent	66 percent
Profit Margin (annual average)	5.5 percent	22 percent
Return on Common Equity (annual average)	7.6 percent	13.1 percent

Source: Bell Atlantic and AT&T SEC Form 10-K, various years

monopoly local telephony market provided steady cash flow to invest in digital switches that formed the backbone of an "intelligent network." The digital system enabled the sale of unregulated, higher-margin custom calling features while also opening the door to other unregulated information services. In 1992, the FCC gave Bell Atlantic and the other RBOCs cellular licenses in their wireline footprint. Bell Atlantic revenue grew substantially over the next decade, with profit margins and shareholder return far exceeding those of the much larger AT&T.[21]

The years following divestiture were a dynamic period of technological change as companies invested tens of billions of dollars in digital switches, fiber optic cable, advanced network software, and highly sophisticated signaling systems. These emerging intelligent networks integrated computer capabilities with communications networks, creating new information services for consumers and businesses and greatly expanding capacity for data transmission. FCC regulators, in a series of *Computer* proceedings, deregulated information services and mandated open architecture networks to spur third-party development of equipment and new software applications that would transform the public switched telephone network into what would become the internet.[22]

The conversion from analog to digital networks also led to the revolutionary transformation of the pre-divestiture small customer service business offices into large, centralized call centers. The on-site private branch exchange (PBX) digital switch running advanced automatic call distributor (ACD) software had the capacity to identify and route calls to the next available agent across multiple call centers. The ACD software gave managers detailed reports on call volume, call length, and customer wait times. Managers used this information to fine-tune service representatives' schedules, minimizing any downtime between calls, effectively speeding up the work process. According to a 1987 U.S. Office of Technology Assessment (OTA) report, "the (automatic call) distributor (ACD) is automatically recording the type of call, the time the call arrived, the identity of the employee to whom it was routed, the number of seconds before the employee picked up, the time the call started, the time the call ended, the number of times the caller was put on hold and for how long, the extension to which the call is transferred, the number of seconds before that person picked up, and so on. In addition, it can show the supervisor at any moment which operators are busy, which are waiting for work, which are on break."[23]

Beginning in the early 1990s, AT&T marketed a software product called "Telemarketing Operations Performance Management System" (TOPMS) that, according to its promotional brochure, "provides you (the call center manager) with a real-time picture of call center and agent productivity. ... Every 3 to 15 seconds, TOPMS gathers call load and agent productivity statistics from your call centers, then sends that data to a command-center workstation."[24] The introduction of interactive voice response systems (IVRs) in the mid-1990s further automated the call center process by enlisting the caller in the call routing process, and at AT&T, eliminating any interaction with a human being for low-revenue customers. These automated systems (press 1 for new services, press 2 for billing) frequently misdirected customers, who took their frustration out on the service representative. Calming irate customers placed high demands on her emotional skills, raising stress levels as she simultaneously struggled to meet her time management and sales benchmarks with no adjustment for the impact of the malfunctioning call routing systems.

CWA's Hobbesian Choice at AT&T

After divestiture, AT&T had to set up an entirely new consumer and small-business customer care operation to serve 80 million residential and 7 million small-business customers as it took over these functions from the local telephone companies. Before the breakup, AT&T's commercial marketing department was limited to several thousand service representatives who assisted marketing managers with large-business customers.[25] Over the next decade, the number of frontline non-management customer service employees at AT&T grew to more than 13,000 as the company opened, reorganized, closed, and consolidated call centers.[26] While some of these employees transferred from local Bell companies or AT&T operator services, most were new hires. About two-thirds were women. While no data is available on racial breakdown, workers recall racial diversity in the AT&T consumer centers located in metropolitan areas with significant African American or Hispanic populations.[27]

AT&T organized its hundreds of new call centers along functional lines based on the type of customer (residential, small, medium, or large business) and type of service (service orders and sales, billing inquiry, credit policies, telemarketing).[28] In the consumer and billing offices where service representatives dealt with low-revenue customers, AT&T prioritized cost control and deployed the automated systems to speed up the labor process and impose strict work time benchmarks. In the early years, employees who served large- and medium-sized business customers had greater autonomy and flexibility, and less supervisory monitoring, but over time, as Mazzeo explained in the opening to this chapter, AT&T moved the dividing line down as it imposed greater surveillance and job pressures in these call centers as well.[29]

AT&T deployed the capabilities of its national network and call distribution technology to consolidate its customer service operations. Soon after divestiture, AT&T opened twenty-four consumer sales and service centers (CSSCs), but later consolidated operations into a nationwide call queue, reducing the number of centers to thirteen in 1993 and to seven mega centers in 1995.[30] In the mid-1990s, AT&T consolidated twenty-four business customer billing operations into just four locations.[31] Consolidation forced employees who worked at the shuttered centers and wanted to keep their union jobs (and seniority) to move, sometimes multiple times. The consumer centers grew to as many as 4,000 frontline workers (plus managers) in one location, some operating twenty-four hours a day. Workers no longer had a feeling of family in these massive mega centers, while managers found it more difficult to provide personalized oversight and relied more on the computer-generated call handling data. "As a call center manager," Jan Schmitz, a Lee's Summit, Missouri, mega center manager, explained, "you've got to make the reps feel they are doing this for themselves. The only way to do this is to build comradery, know the person, how doing a good job will help them. With big centers and lots of turnover you don't have personal attention. There was more management by the numbers at Lee's Summit (mega center)."[32] The workstations in the new call centers were designed to isolate employees, to focus the representative's attention on the computer screen and the customer. The cubicles were arranged in long rows with dividers between each cubicle. Employees had to peer over or around their cubicle to talk to each other. This made it more difficult for representatives to help each other and to chat and make friends.[33]

In taking over customer service operations from the local Bell companies, AT&T considered outsourcing its consumer operation to a third party such as American Airlines or Electronic Data Systems (EDS), companies with the most sophisticated automatic call distribution systems at that time. In July 1984, as AT&T began setting up its consumer customer service division, labor relations director Robert Livingston reached out to James

Irvine, CWA vice president with responsibility for the AT&T bargaining unit. Livingston explained that AT&T would not pay the rates currently paid to AT&T customer service representatives and other clerical titles in the new centers. He offered a carrot and a stick: if the union agreed to create a new account representative and other clerical titles at a 20-percent lower wage rate, AT&T would create thousands of union-represented call center jobs. But if the union refused, AT&T would outsource the work. For the account representative, a 20-percent wage cut reduced her annual earnings from $23,751 to $19,000, an annual loss of $4,751. This demand was unprecedented in the Bell system. Never before had AT&T or any Bell Operating Company (BOC) approached the union with such an ultimatum. AT&T claimed that the wage cut was necessary to compete with the lower labor costs of the non-union long-distance companies such as MCI and Sprint. With this demand, the new AT&T made it clear that driving down labor costs was a key component of its competitive strategy.[34]

AT&T's ultimatum presented CWA with a painful choice—one that has become increasingly common in labor negotiations, beginning in the concessionary bargaining era of the 1980s and continuing today, as unionized companies demand wage concessions or second-tier wage scales under the threat of plant or office closures and layoffs. Because this was new work, CWA had no contractual method to prevent AT&T from outsourcing the jobs. Irvine concluded that the union had to take the wage cut. He thought he would receive limited political pushback from the membership, since the affected jobs were not yet populated. As a CWA official explained years later, "Without recounting all the discussions and soul-searching we endured to reach a decision . . . the basic reasoning was that it would be easier to increase the wages of employees in the bargaining unit than it would be to organize the employees in companies contracting the work from AT&T, then getting a first contract, then increasing the wages."[35]

In August 1984, Irvine signed a memorandum of agreement establishing the new job titles and wage rates. The agreement proved to be extremely controversial with CWA activists and local presidents, who mobilized in every subsequent bargaining round to eliminate what they considered "second tier" titles. Irvine defended his decision, noting first that the account representatives would be limited to answering billing questions, with no sales responsibilities, and second that AT&T told him the company would create thousands of new union jobs as a result of the agreement. The number of jobs did increase; by 1995, more than 8,700 employees were working at the second-tier rates.[36]

In the next four triennial rounds of contract negotiations (1986, 1989, 1992, and 1995), the union prioritized upgrading these titles in its list of

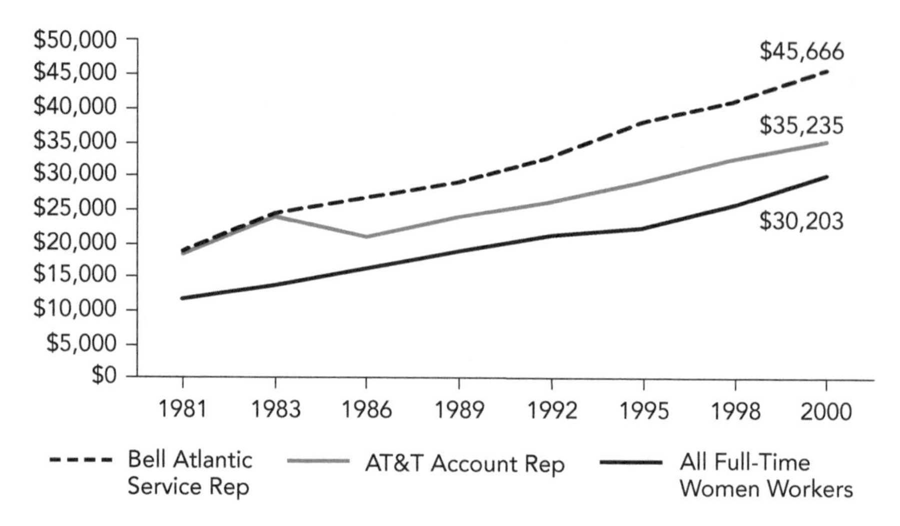

Annual Earnings of Bell Atlantic Services Representatives, AT&T Account Representatives, and All Full-Time Women Workers, 1981–2000.

bargaining demands. Each year, the union succeeded in getting the account representatives a small wage increase (over the general wage increase) but was unable to win its demand to eliminate the salary gap. In 1989, CWA did achieve an upgrade to a new title with an 8.5-percent to 10-percent wage increase for about 1,300 account representatives serving business customers. The higher wage rate was designed to address high rates of turnover, as many as two-thirds of new hires were leaving the job due to low pay and high stress. In 1998 negotiations, unable to resolve several controversial wage design matters, including the second-tier titles, AT&T agreed to allocate $10 million to a joint union-company committee to address the issues. CWA proposed moving all second-tier titles to the higher customer representative wage schedule; after AT&T refused, CWA took the case to arbitration. The arbitrator decided to award the second-tier title recipients a 2-percent wage increase, effectively preserving the second-tier pay grade.[37]

Thus, despite the union's efforts over five separate contract negotiations, in the year 2000, an AT&T account representative earned about $35,000 per year, which was $10,000 less than the average $45,000 annual earnings of a Bell Atlantic service representative. The cumulative difference for an AT&T account representative who had worked every year from 1984 through 2000 would amount to over $100,000. To be sure, an AT&T account representative's annual earnings were about 14 percent higher than the median for all U.S. full-time female workers, and the AT&T

employees had superior health, pension, and other benefits, as well as union protection on the job. Although the account representatives fought hard to elevate their issue on the union's bargaining agenda, the CWA leadership prioritized stemming the massive job cuts of Western Electric factory workers, technicians, and operators who formed the power base of the union. Eliminating the second-tier titles was simply not a top priority and certainly not a strike issue.

Blocking Job Downgrades at Bell Atlantic

CWA units at Bell Atlantic faced fewer competitive market conditions and proved more effective in blocking management attempts to downgrade the service representative job. Still, the company responded to divestiture with a laser focus on labor cost containment. Unlike AT&T's greenfield operations, Bell Atlantic did not start fresh in designing centers after divestiture. Bell Atlantic's service representatives had worked under union contract for decades; the company knew that union power would block any direct assault on union wage rates. Rather, to boost profits in the residential market, where most service representatives worked, Bell Atlantic set out to transform service representatives into salespeople; reorganize work to give certain functions to lower-paid employees; and cut costs by pushing service representatives to serve more customers in less time, even as the job became more complex.

Bell Atlantic's core business of local telephony was a mature market with low profit margins. Divestiture put an end to cross subsidies from long-distance rates, while state regulators kept local rates low even as the company faced significant capital expenses in the conversion from analog to digital switches and investment in fiber trunk lines. Bell Atlantic, therefore, sought to boost return on its telephony business by selling bundled services to residential and business customers, and dedicated switching services (known as Centrex) to business customers, and above all, reducing labor costs through job cuts, automation, and work process efficiencies.

Before 1986 contract negotiations, Bell Atlantic labor relations staff delivered to CWA negotiators a document entitled "Bargaining '86—Preparing for Tomorrow" detailing how the company's post-divestiture business strategy would drive its labor relations. "To survive in this new environment," Bell Atlantic explained to CWA, "we must have a competitive cost structure and sufficient flexibility to respond to ever changing business conditions." In the two years since divestiture, the company had already reduced the workforce by more than 10,000 employees—half management, half union-represented employees. Going forward, the company said, it would need to cut more jobs and pursue other "cost containment efforts, combined with the need to create compensation systems more appropriate to a competitive business" and "increase productivity by selective implementation of mechanization and by strategic use of our force at all times." To be sure, Bell Atlantic labor relations staff wrote the "Bargaining '86" document to dampen CWA expectations in contract negotiations. The company was sending CWA a powerful message: say goodbye to the steady improvements in wages, benefits, and working conditions that the union and its members had come to expect during the monopoly era. In effect, the company used the shift in market conditions to destabilize traditional labor-management relations.[38]

The document also reveals a tension in the way Bell Atlantic viewed its customer service operation—was it a cost center or a value creator? The service representatives believed they were value creators whose compensation should reflect their worth. But Bell Atlantic managers reached a different conclusion, at least regarding the vast majority of Bell Atlantic service representatives who spent their time assisting low-margin residential customers. In management's view, it was only when basic dial tone was bundled with unregulated custom calling services that the residential customer would become a profit center. Therefore, Bell Atlantic set out to reorganize the division of labor by giving service functions to lower-paid employees and using automated time management systems to push service representatives to serve more customers in less time.

In response, service representatives working for Bell Atlantic's C&P Telephone subsidiary (operating in Maryland, Virginia, West Virginia, and Washington, DC) mobilized before the 1986 round of collective bargaining, the first since divestiture, to pressure the negotiating team to demand a higher wage rate for their job title, one that would reflect, as one service representative put it, "the changes of job responsibilities and the constant increase in duties." The service representatives defended their demand for a wage increase. "We service reps are the front door to C&P. We are the #1 Revenue-producing segment of our company." These workers wanted to make clear to their union—and to Bell Atlantic—their vital position within the company and their potential power in the upcoming negotiations.[39]

Judy Buchanan, whom we met at the beginning of this chapter, together with thirteen other service representatives, submitted to CWA negotiators a ten-page description of the job duties they performed as billing representatives. Although their primary responsibility involved billing and collections, these service representatives were now required to sell additional services on every call. Even if a customer called to complain about a bill, the billing representative was supposed to make a sales offer. This required toggling between two different computer screens on 80 percent of customer contacts (the toggling took a precious forty-five seconds, counted against her average talk time). She was required to make the sales attempt in addition to her numerous billing duties, which included processing notices of interruption, denial, or restoration of service on 200 to 300 accounts per day. As a result of these job pressures, they explained, "the real drug problem in our office are tranquilizers, pain relievers for headaches and migraines, and tagamet for stomach ulcers."[40]

Service representatives taking service orders and making sales were required to complete fifty to sixty customer contacts per day in three to three and a half minutes per call, while "overlapping" to complete paperwork while the customer was still on the line. With the introduction of computerized customer records, the service representative typed orders into the system—a function previously handed off to service order typists. The company imposed unrealistic sales quotas designed to prioritize sales over service. "Sales is the most emphasized part of our job," one service representative explained. "We are expected to sell or at least attempt to sell revenue items on 100% of our calls. We are expected to meet 150% of our sales objective to be rated 'outstanding' in sales." The service representatives were especially offended by the requirement to follow a predetermined script in conversations with customers. "How can I use the creative part of my brain when YOU insist on developing it for me?" Francis Randall, a service representative in Norfolk News, Virginia, wrote in a letter to her managers that she shared with the union. "I have 'management' reminding me that 'CHT' [contact holding time] is this or that number of minutes or 'APBs' [customer accessibility] at this percentage. Our contacts are growing longer and longer." In another handwritten plea to the CWA bargainers, twenty-eight service representatives summed up their demands: "Service Reps are the liaison between the company and the customer. We have more on-job-stress than any other job title. The job pressures, duties and responsibilities of the Rep continues to increase. . . . It is time to increase our PAY!"[41]

In 1986, the three CWA bargaining units at Bell Atlantic (C&P, New Jersey Bell, and Pennsylvania Bell) won a $5 weekly pay increase for service representatives over and above the general wage increase. At New Jersey Bell, the agreement included an important protection. Recognizing the union's concern that "the complexity of this work does not lend itself to the business office environment of fast schedules, observations, etc.," the company agreed that it would not discipline a service representative rated unsatisfactory in sales, provided the person had worked in the call center for at least six months. This provision would be renewed in subsequent CWA/New Jersey Bell contracts.[42]

Bell Atlantic continued to seek ways to reduce labor costs in its call centers, and in 1989, approached CWA with a proposal to establish a new collector title to handle billing and collections issues, with a wage rate about 10 percent below that of service representative. The CWA bargaining units accepted the proposal. While this functionalized division of labor produced some labor cost savings for Bell Atlantic, it raised new problems. Because Bell Atlantic did not want to lose any opportunity to sell its value-added services, it continued to require collectors to make sales attempts on collections calls. The union filed multiple work-out-of-title grievances, with a significant number escalated to the time-consuming and costly arbitration process.[43]

Bell Atlantic discovered that a strict division between sales and service functions proved unworkable. The company wanted all those answering service calls to bridge to sales. And the company wanted those who handled sales to be able to provide one-stop service, including answering ancillary billing questions. In addition, the company wanted to create regional routing of calls, cutting across the various bargaining unit jurisdictional lines. To achieve these goals, in early 1994, the three CWA bargaining units and Bell Atlantic signed the "consultant letter of agreement" that effectively re-integrated sales and service functions into one job title. The collectors were upgraded to the former service representative, now renamed consultant, rate of pay. Most significant, the agreement included some protections against discipline for failure to meet sales quotas. The letter of agreement stated that "sales results will be a job requirement for the Consultant which specialize in sales, provided that sales results will not be the *sole* basis for discipline" (italics added). In exchange for upgrading the collector title, the union agreed that the company could create region-wide call queues as long as it did not lead to any consultants being laid off, downgraded, or forced into part time. This agreement gave Bell Atlantic the flexibility it wanted to send calls to any available consultant in the five-state and Washington, DC, area, while giving the union-represented collections representatives a 10-percent wage increase.[44]

In Pennsylvania, the service representatives won a significant victory against company-imposed sales pressures through an alliance with consumer advocates in a case before the Pennsylvania Public Utilities Commission (PUC). The case involved a complaint by the Pennsylvania Office of Consumer Advocate against Bell Atlantic for consumer fraud related to its policy that required service representatives to promote more costly service bundle packages over the less expensive stand-alone telephone service.

CWA-represented service representatives, under whistleblower protection, provided powerful testimony describing how company sales quotas drove this deceptive sales practice. In 1989, the company, the PUC, and the consumer advocate settled the case with a mandate that required customer service employees to provide numerous disclosures to customers. Further, to protect frontline service representatives, the agreement prohibited Bell Atlantic from giving a customer service employee an unsatisfactory rating solely for failure to meet sales objectives. These provisions have served to protect Bell of Pennsylvania service representatives to this day.[45]

Several factors explain why the CWA bargaining units at Bell Atlantic were more successful than the CWA bargaining unit at AT&T in protecting, and even slightly upgrading, customer service employees' compensation levels. First, Bell Atlantic operated in a more protected, regulated, and stable market structure. The company simply did not face the same competitive pressures and therefore business imperative as AT&T to reduce labor costs in the call centers. Second, Bell Atlantic's state regulators monitored the company's sales and service practices, which helped ensure that sales practices both were ethical and did not eclipse service functions, whereas the FCC as AT&T's regulator largely saw competitive choice as the best route to consumer protection. Third, CWA had more power in the Bell Atlantic bargaining units to influence service representative issues. The CWA bargaining unit at Pennsylvania Bell was composed exclusively of customer service workers, and the New Jersey Bell unit included only customer service employees and operators. The female union leaders in these units were or had been service representatives themselves, with deep knowledge of and passion for their issues, and had developed long-standing relationships with their company labor relations counterparts. Their political base was composed solely of customer service workers (and in New Jersey, operators). In contrast, customer service employees represented only a small portion of the nationwide CWA bargaining unit at AT&T and many were relatively new to the job and to the union. At AT&T, the CWA negotiators came from technician or operator ranks, which was their political base, and stanching the hemorrhaging of those jobs, rather than addressing service representative issues, was their top priority in contract negotiations.

Organizing for Power: The Commercial/Marketing Conference

During the decade after divestiture, customer service workers organized to build greater power within CWA to elevate their issues on the union agenda. As business changes in the telecommunications sector drove growth in the number of service representatives at the same time that conditions deteriorated in the call centers, customer service workers mobilized to push CWA to give greater weight to their concerns. They built power through various mechanisms: signing up more members, running for union office and bargaining committees, networking at CWA regional and national meetings and those of the Coalition of Labor Union Women (CLUW), developing relationships with leaders and staff, organizing coworkers to submit and push their bargaining demands, and seeking staff appointments. Most significant, CWA customer service workers took a cue from the CWA operator activists who met annually to network and elevate their issues within the male-dominated technician union leadership. The CWA customer service members organized their own annual commercial/marketing conference, bringing together hundreds of CWA customer service activists.

At AT&T, the first union challenge was to organize the workers at the new call centers. While the union had a contractual right to represent service representatives, it had to sign them up. AT&T did not interfere as CWA technician locals discovered the new call centers in their geographic jurisdictions and proceeded to enroll the employees as members and build steward structures. AT&T's neutrality was noteworthy, especially in contrast to its aggressive opposition to CWA representation in its new credit card, telemarketing, and computer subsidiaries. In the call centers, AT&T managers abided by the company's commitment to CWA to grow union call center jobs in return for the second-tier wage agreement. Moreover, the company was not willing to violate the contractual recognition clause in its core communications business, still a powerful CWA power base within the corporate family.

Glen Hamm was a technician and union leader working at a major AT&T network operations center outside of Chicago. "One day there was nobody [in the Oakbrook, Illinois, call center] and the next day there were maybe two or three hundred people," he recalled. "My local asserted jurisdiction and began to sign them up as members." Mary Ann Alt has similar memories of the unionization process at the Account Inquiry Center in Hunt Valley, Maryland. "Jerry Klimm of local 2150 was a technician. He came and signed us into the union." AT&T did not oppose these unionizing efforts.[46]

The technicians, with their more independent craft traditions, were appalled at the working conditions in the call centers. "I was hearing horror stories," Hamm remembered. "The service representatives were constantly monitored. They had to raise their hands to go to the restroom. It sounded like AT&T was bringing these people on and treating them like operators in the 1930s." Hamm invited local presidents (mostly male technicians) who represented AT&T call center workers, as well as emerging union activists who worked in the call centers, to a conference at a Chicago airport hotel. In the early years, he also invited AT&T managers, reflecting a belief that the union and management could partner to improve working conditions. Within a few years, the call center activists, predominantly women, took over leadership of the CWA commercial/marketing conference, and the AT&T managers were no longer invited.

The annual gathering was an opportunity for union leaders and activists from the call centers to network with each other, develop leadership skills, and press the male negotiators to take up their concerns. "We'd have the bargaining committee at the front of the room and we would have a microphone for the participants to present what we thought were the most important issues to ask and address in bargaining," Alt explained. "We began to learn how to prepare bargaining demands." Eventually, many of these women were elected to top leadership positions in their local unions and the designated commercial/marketing slot on the AT&T national bargaining committee. Moreover, the size of the call centers led to large locals, which in turn led to more political power and attention from regional and national union officials.[47]

In 1991, the CWA commercial/marketing conference expanded to include activists from every RBOC bargaining unit. This was a bottoms-up gathering, organized by a team of local leaders from across the country, with local unions footing the bill for expenses of the hundreds of stewards and local officers who attended. The CWA activists invited regional vice presidents, staff, and national leaders to these meetings to learn about service representative issues, and so the activists could impress on them the growing importance of this segment of the union. CWA Executive Vice President Dina Beaumont (whom we met in chapter one) represented the national union at the conference. Beaumont was a former operator who not only understood the job pressures of a highly regimented, surveilled work environment, but was also a savvy political operative. She understood that as automation decimated the operator workforce, the future of female membership and leadership in the union depended on the growing customer service segment. She nurtured a network of service representative leaders who shared information about the latest developments in the call centers and who strategized with each other to develop solutions. The information exchange served as an early warning system alerting union leaders to new developments, and as a method to fact-check employers' claims about what the union had agreed to in other locations.[48]

The Fight Over Surveillance

The Bell system had a highly developed and tiered system of supervisors who tracked worker performance to maximize efficient operations and conformance to the detailed methods and procedures. In an automated call center, the digital equipment gives managers unsurpassed power to track their workers' job performance and to use that information to speed up and deskill work.[49] As discussed in chapter 1, Bell system telephone operators responded to the introduction of electronic surveillance with a job pressures campaign culminating in 1980 with AT&T, and then RBOC, contract language that banned secret monitoring. The provision required supervisors to sit beside the operator during any observation sessions, effectively ending the use of secret monitoring to "catch" an employee. However, the contract language only covered operators. It did not apply to what was then a much smaller group of service representatives who, at the time the language was first negotiated, had more autonomy on the job, were more likely to perceive supervisory monitoring as a training tool, and were only peripherally involved in the CWA monitoring campaign.

In the post-divestiture decade, AT&T and RBOC supervisors intensified surveillance of customer service employees. Management deployed two major forms of supervisory oversight: first, through the more traditional method of listening in on the service representatives' conversations with customers, and second, beginning in the early 1990s, digital tracking of the service representatives' time on the phone, including average talk time, time between calls, and most important, adherence to one's assigned schedules. The service representatives deeply resented the use of monitoring for punitive purposes, a sign that their supervisors saw them as children prone to misbehavior rather than as professionals dedicated to serving their customers. They particularly resented the automated management systems that treated them as numbers on a computer printout rather than as skilled workers, and they hated the work speed-up and constant surveillance that turned their workplaces into hothouses of stress. "I despise how I am devalued, degraded, and humiliated on my job," one service representative wrote on a 1997 CWA stress survey. "Everything is measured and monitored. There is no fun at work, no laughter. There is not a day that goes by that someone does not break out in tears." Another added: "The only thing that matters now is the numbers that flow from the monitoring equipment. We are like factory workers on an assembly line. I used to love my job but now I hate it."[50]

The union made some progress placing limits on secret supervisory observations in the business offices. Building on the operator precedents,

CWA in 1992 negotiated strong protections against secret supervisory monitoring at AT&T, extending the operator language to service representatives and adding a prohibition against using monitoring as the sole basis for discipline. Service representatives and union leaders remember that the contract language largely protected employees from abusive supervisory monitoring in the AT&T consumer call centers. Bell Atlantic proved much more resistant to the union's monitoring demands. It was only after a major mobilization of CWA-represented consultants and a two-week strike in the year 2000 (as discussed in chapter 6) that Bell Atlantic agreed to a comprehensive stress relief package for consultants, including annual limits on the number of monitoring sessions.[51]

While the companies made some concessions to limit traditional supervisory observations, they were unwilling to cede any control over the digital time management systems. Beginning in the early 1990s, AT&T, Bell Atlantic, and the other RBOCs took advantage of the increasing power of digital technologies to develop workforce management software to track and forecast call volumes in fifteen-minute increments, and then to schedule the exact number of service representatives needed to meet those call volumes, squeezing out "non-productive" time off the phones. The result was an enormous speed-up of the work, eliminating any free moments between calls to do paperwork, chat with a colleague, get a drink of water, go to the bathroom, or simply get a break from a difficult customer interaction.

To make the time management system work, the companies adopted what were called strict adherence policies that tracked whether the service representative was following her assigned schedule, measuring any deviation down to the second. Supervisors watched a screen with wavering color-coded bar graphs, each representing a service representative, tracking each employee's time on or off the phone and whether this conformed to her assigned schedule. Supervisors could see this data in real time and in daily printouts for each person, work group, and call center, and all centers on the same call distribution channel. While this workforce management system was a remarkable technological achievement, the companies used it to adopt strict adherence policies that the service representatives deeply resented, not only because of the pressure but also because the policies interfered with their ability to provide good service.

The CWA leadership learned about the new adherence policies at the 1995 annual commercial/marketing conference. Beaumont encouraged conference participants to pass a resolution calling on the CWA executive board "to establish a National Task Force, consisting of local leaders with Commercial Marketing experience from each [CWA regional] district to

Adhere This, Adhere This
Big Brother is Watching You

Report of the CWA
National Task Force on Adherence

Communications Workers of America
May 1995

CWA Adherence report cover, CWA research department, 1995.

survey, investigate and report on the problems and activities of the Bargaining Units on the issues of ADHERENCE." The CWA executive board heard their call and created the national adherence task force, chaired by Beaumont and composed of CWA customer service leaders from the various bargaining units. The group convened for two days in Washington, DC, and then issued their report, *Adhere This, Adhere This: Big Brother Is Watching You.* The report cover featured a female detective with a spyglass, conveying the report's conclusion that "adherence policies raise employee monitoring to a new level of indignity."[52]

The report included this description from an AT&T account representative of how adherence set up a catch-22 for her. "If I'm supposed to take a break at 10:30 but I'm on line finishing a call with a customer until 10:32, I'm out of adherence. If I take my 15 minute break and return at 10:47, I'm out of adherence again. If I go off-line to another department with a customer inquiry, I'm out of adherence. If I take too long with customers, I'm not meeting average call time and I'm out of adherence." Adherence deviations included staying on the line with a customer when one is supposed to be on break or at lunch; closing the phone to incoming calls to reach out to another department to solve a problem or do a few minutes of paperwork; taking too long to log into the system; going to the bathroom or getting a drink of water off schedule; and taking a sick day or an excused personal leave day after the schedule has been issued. The adherence pressure was most intense in the residential sales and collections offices, with adherence set at 95 percent at AT&T, 92 percent in the Bell Atlantic sales centers, and 85 percent in the Bell Atlantic collections centers. Bell Atlantic policy warned that any violation of the schedule "may result in disciplinary action, up to and including dismissal on the first offense."[53]

The task force report identified multiple problems with adherence: inaccurate computers, favoritism in enforcement, inflexible systems, increased stress-related illnesses, impediments to good customer service, incentives to game the system, and dehumanizing invasion of privacy, dignity, and autonomy. "In my office of 38 reps, 5 are out on stress-related disability," reported a Bell Atlantic service representative from Reading, Pennsylvania. "Three more have colitis and another has an ulcer. . . . Not a day goes by that I don't take a Motrin for headaches." The report emphasized the irony that the companies were "implementing these inhumane policies at the same time that they talk about restructuring work to 'empower' employees to make decisions. It is impossible to empower employees to solve problems when big brother tracks every second of one's day and punishes for any deviations."[54]

In May 1995, the CWA executive board adopted a CWA policy on adherence and sales quotas. "Adherence equates to electronic stalking of employees," the policy began. "Adherence forces employees to sacrifice their health and human dignity in order to keep their jobs." CWA articulated an alternative vision. "In a competitive environment, the winners will be those employers who treat their employees with dignity, autonomy and respect and who provide customers with excellent customer service. Experienced and motivated employees are the key to competitive advantage. CWA knows that employees want to do a good job and, given proper training, supports, and staffing levels, will provide good customer service

Adherence: "Sticking It" to the Worker.

I'M SORRY, SIR. BUT DUE TO ADHERENCE, MONITORING AND SALES QUOTAS, I HAVE NO TIME IN MY DAILY SCHEDULE TO HELP YOU.

CWA flyer distributed as part of national mobilization against abusive adherence policies, CWA research department, 1995.

without the need for monitoring, adherence, or sales quota speed-up." The policy recommended bargaining proposals to limit the abusive impact of adherence and sales quotas.[55]

CWA customer service locals across the country mobilized for two days in the summer of 1995 around adherence. The national office distributed a flyer for reproduction by locals entitled "Adherence: Sticking It to the Worker." The flyer included a picture of a service representative at her desk, chained to a clock on her back, saying to her customer: "I'm sorry sir. But due to adherence, monitoring, and sales quotas, I have no time in my daily schedule to help you."[56]

Service representatives across the country participated in the national adherence mobilization. They wore "adhere this" T-shirts and buttons, with pictures of a service representative in ball and chains. They handcuffed themselves to their desks. They rallied outside their employers' downtown corporate headquarters, passing out flyers. One local in the midwestern Ameritech region produced a coloring book, including a picture of a male supervisor handing out catheters while telling a group of service representatives that "I can't make you wear these, but it sure would help adherence."[57]

Despite the mobilization, CWA negotiators made little progress curtailing their employers' strict adherence policies in the 1995 and 1998 rounds of bargaining. For the companies, negotiating with the union over time

Page from CWA midwestern district 4 coloring book distributed as part of mobilization against adherence policies, CWA research department, 1995.

management and adherence policies threatened the very essence of managerial control in the call centers. Supervisors would not cede authority over how their employees spent their time on the job. Only through strict adherence to schedule would management's efficiency schemes function properly. The fight over adherence, even more than monitoring, was at its very foundation a struggle over power to control the labor process in the call center, which management refused to cede.

Struggles over Sales Quotas and Sales Commission Plans

As sales became a more important component of the service representative job, management at AT&T and Bell Atlantic approached CWA with proposals to introduce a sales commission component to the compensation package. Sales commissions, unlike base wages, compensate employees based on sales results; the more an employee sells, the more money she can earn. Commission plans can be structured in various ways. Piece rate systems compensate an employee for each product sold. More typical sales commission plans in call centers establish a sliding scale payout that increases as one sells more, often requiring the sales representative to meet a minimum sales benchmark before any payout is available. The commission plan usually includes various components, including total revenue sold, sale of particular products, and quality measures. Compensation consultants advise companies in the complex design of sales commission plans. According to the American Compensation Association, plan design should set objectives at a level that would allow about two-thirds of the work group to achieve the target, while encouraging the top 10 percent to 15 percent of sellers to exceed the quota. At least 90 percent of the sales force should receive some payout, leaving about 10 percent without any sales commission compensation.[58]

Because compensation is a mandatory subject of bargaining, AT&T, Bell Atlantic, and other RBOCs were required to negotiate with CWA over introducing a commission plan. The negotiations over these plans, and, where CWA agreed to them, the struggles over their implementation, reveal the widely divergent value systems of management and the CWA over what constitutes fair compensation. Company managers placed highest priority on revenue generation; in a fair system, they believed, workers who sold more should reap larger monetary rewards. In contrast, union leaders placed highest value on worker equity, solidarity, and compensation that provided the service representative with a fair share of the value she created for her employer. The union responded with great skepticism, if not outright opposition, to any management proposal to put a portion of the base wage "at risk" based on sales results. In the three cases I discuss below, we see this clash of values, even as CWA units reached different conclusions and evidenced different power relations in response to company demands for sales commission plans.

In 1992 contract negotiations, AT&T approached CWA with a proposal for a commission plan for the sales representatives working in the call centers serving small and midsize businesses. The proposal was an "80/20" plan, meaning that the sales representative would receive a base wage set at 80 percent of the current wage rate and the other 20 percent would be earned if she made 100 percent of her sales target. Initially, the elected bargaining team, with CWA Local 1150 President Laura Unger representing the customer service work group, rejected the proposal. Unable to reach agreement with the company on economic issues, the CWA national bargaining team, led by Vice President Jim Irvine, recessed all negotiating tables. The CWA team that had been negotiating the commission plan gave Irvine a list of ten provisions that were essential to protect employees if the union were to agree to a commission plan. At a union rally in Syracuse, New York, Unger, standing on a flatbed truck in front of the AT&T call center, made it clear to the sales representatives that a forced commission plan was a strike issue.

Yet, once the union reached agreement with the company on an economic package, Irvine accepted the company demand for an 80/20 commission plan. The bargaining team objected to the decision by the national union leadership to take the rest of their demands regarding the commission plan off the table. The bargaining team refused to sign the contract (a purely symbolic move) and urged the membership to vote down the contract. But since the commission plan affected only a small percentage of members, they were not able to win majority support for a no vote, and the contract won approval.[59]

Initially, the commission plan was lucrative, and most sales representatives were able to earn their "at risk" money and even more. In 1995 bargaining, the union won a small improvement in the plan: quota relief for those on vacation and other contractually guaranteed time off the job. The union also tried to fix another problem. The company required sales representatives to meet two different sets of sales benchmarks, one for compensation payouts under the commission plan and the other to drive sales promotions of products. Sales representatives who made their sales commission targets were getting disciplined, including termination, for failure to meet the quotas established for product promotions. After lengthy conversations, CWA believed that AT&T negotiators made a verbal commitment that employees who met commission plan quotas would not be disciplined for failure to meet promotion plan quotas. However, AT&T managers continued to discipline workers who missed the promotion plan benchmarks. In 1999, CWA brought the issue to arbitration. The arbitrator ruled against the union, noting that the 1995 negotiated quota relief was designed to address the issue. The arbitrator missed the point entirely. After that, AT&T supervisors continued to discipline employees who missed their sales benchmarks, and as Mazzeo described at the beginning of the chapter, many of her coworkers dreaded going to work for fear they would not make their sales quotas and would face discipline or termination.[60]

In contrast, at Bell Atlantic, the union had the power to squash management's proposal in 1995 contract negotiations for an 80/20 sales commission plan (80 percent base wage, 20 percent "at risk" based on sales results). The union rejected the proposal after reviewing company sales performance data from the prior two years. In 1994, almost half (48 percent) and in 1995 almost two-thirds (63 percent) of service representatives missed their individual sales objectives. The data made clear that a company-designed commission plan would reward high sellers, while average and low-volume sellers would likely see a reduction in compensation, and sales pressure would intensify. CWA was not willing to agree to a commission sales plan that would harm the majority of the membership and rejected the company proposal.[61]

CWA negotiators at US West Communications reached a different conclusion in 1995, becoming the first CWA bargaining unit to agree to a commission sales compensation plan at one of the seven RBOCs. US West was the smallest of the RBOCs, with $9.3 billion in revenue in 1995, serving the geographically dispersed fourteen-state region west of the Mississippi River.[62] In contract negotiations, CWA District 7, representing about 5,000 sales consultants working in ten large call centers, proposed a substantial wage increase for this group. US West countered with an 80/20 leveraged compensation proposal, their top objective in the negotiations. To sweeten the deal, US West offered to set the annual guaranteed base at $32,000, which was then the wage rate for sales consultants. The company had sole discretion over plan design, while a union-management advisory committee would monitor the plan and review attainment levels. "Objectives set for each performance level will be stretched—but attainable," company negotiators explained. CWA accepted the proposal because, as Annie Hill, administrative assistant to the District 7 vice president and bargaining committee chair, recollected, "This was a win-win for the company and our members. With the guaranteed wage base, we thought people would be able to make more money."[63]

The first year of the program, according to Hill, was the "honeymoon period, people were making money hand over fist." But by the second year, the company raised the objectives, service representatives were not making back "at risk" money, and many were getting disciplined or terminated, were asking for a transfer, or were leaving the company due to high stress levels. Forty-one percent of sales consultants in consumer and 38 percent in small business were not making their individual sales revenue objectives. Overall commission payout dropped 21 percent in consumer and 34 percent in small business from the prior year.[64]

In 1998 contract negotiations, CWA made fixing the objectives its top priority and presented the company with a demand to set the objectives so that at least 90 percent of the work group would attain the target. The company refused. The company negotiator, Ginger DeReus, presented the union with an opinion letter from compensation consultant Hewitt Associates. "It's been a long-standing rule of thumb in the design of sales compensation programs," Hewitt wrote, "that, on average, about two-thirds of the sales force should achieve quota in a typical year." CWA and US West were unable to agree on a specific attainment objective and therefore, to resolve the issue, inserted a contractual requirement that the plan objectives be "reasonable," effective January 1, 1999.[65]

Despite the employer's contractual obligation, the CWA oversight committee review of commission data found a "disturbing trend" of low sales consultant attainment levels. The call sharing groups (the work groups that shared the same call distribution queue) attained the two-thirds "rule of thumb" only 18 percent of the time in consumer and only 23 percent of the time in small business. CWA filed a grievance, which led to an arbitration over whether the sales objectives met the contractual "reasonableness" standard. In several days of hearings before an arbitrator, CWA presented statistical evidence and sales consultant testimony to support its case that the sales objectives were "unreasonable," resulting in discipline, stress, and high turnover rates. In the seventeen-month period covered by the arbitration, the "two-thirds" industry standard was met in only two months in consumer and never met in small business. About half of the sales consultants received no payout on the individual revenue component.[66]

The failure to meet individual sales revenue objectives could, and did, result in discipline and termination, accounting for high and expensive turnover rates. According to a US West internal study, the turnover rate among sales consultants ranged from 40 percent to 50 percent, at an annual cost to the company of $52 million. One CWA witness testified at the hearing that 43 percent of all sales consultants in the Phoenix and El Centro, Arizona, call centers that she represented had resigned or transferred to another job—even one with lower wage rates—due to the pressure from unreasonable sales objectives. Another CWA witness reported that only two sales consultants remained in her work group since the inception of the commission plan, and a third CWA witness testified that out of an initial training class of nineteen employees, only two remained.[67]

The CWA witnesses testified to the enormous stress that they experienced because of the unreasonable sales objectives. "It doesn't feel good going to work every day knowing that I'm not succeeding," Sharon Goldberg explained. Unrealistic sales objectives undermined employee morale. Sales pressure and fear of discipline created incentives to engage in the illegal practice of "slamming," or putting items on customers' accounts that they did not order. A former manager in the Des Moines, Iowa, call center testified that when she reported one such code of conduct violation, her supervisor told her to "turn the other way" because the "slammer" was a high sales revenue performer and the call center needed him to meet its team objectives.[68]

In April 2001, arbitrator Harold Wren issued his decision in the case, ruling that the US West sales objectives were unreasonable and in violation of the contract language. However, his remedy reflected a complete misunderstanding of the way the sales compensation plan was structured and in fact made the situation worse. He ordered US West to reduce the payout levels—not the revenue objectives—by 10 percent. His ruling allowed the company to retain the unrealistically high sales objectives while mandating a 10-percent cut in the sales commission earned.

These cases illustrate the slippery slope that challenged the union's efforts to negotiate commission plans. Without a meaningful voice for the union regarding plan design, the company set quotas and payout rates that undermined union values of equity and solidarity, challenged individual workers' sense of self-worth on the job, intensified the pace of work, and created perverse incentives to engage in unethical sales practices while shortchanging service. For these reasons, CWA rejected a sales commission compensation plan at Bell Atlantic. And as CWA learned at AT&T and US West, once the union opened the door to a commission sales plan, it was very difficult to negotiate plan design to ensure that all workers received fair compensation for their work with protections against inordinate stress and speed-up.

Occupational health and safety researchers have found that the most stressful jobs are those that place high demands on employees but give them little control over their work. This describes the work organization in the post-divestiture AT&T and Bell Atlantic call centers. A 1997 CWA survey of its customer service members concluded that there was a "stress epidemic" raging in their workplaces. Two-thirds of the 1,600 workers who responded to the survey reported one or more stress-related illnesses, including migraines, depression, high blood pressure, sleep problems, eating disorders, chest pains, and hives. One-third reported that they had taken time off work in the past year due to stress-related illness; of these, about 10 percent were out more than one month. They described their call centers as an "electronic sweatshop," "kindergarten," or simply "workplace from hell." To be sure, this was not a scientific survey, but even accounting for some hyperbole, the results indicate serious problems.[69]

CWA leaders recognized that competition was driving down their power to negotiate good compensation packages and safe, humane working conditions at the union companies. Gone were the pre-divestiture days of sectoral bargaining, when the union contract set the wages, benefits, and working conditions in the telecommunications industry. Gone also were the days when the union grew as the company grew. The union faced new imperatives to organize the new-entrant non-union companies and to gain union representation of workers at the new subsidiaries that their employers were creating union-free, and to pursue alternative labor-management approaches to union power to improve conditions at work.

CHAPTER 3

Organizing to Block the Low Road Path

"Within three months of arriving at LCF [Sprint Corporation's La Conexion Familiar], I was asked to spy on my co-workers," Liliette Jiron testified at a U.S. Department of Labor forum in San Francisco in February 1996. "I felt that I had no choice. . . . I couldn't lose my job. . . . I understood why my co-workers wanted to form a union. We had problems getting paid. . . . They kept changing the rules on the number of sales we needed. . . . They told us not to drink a lot of water so we wouldn't need the bathroom breaks. . . . They kept telling us if we voted for the Union, the office would close down." Then, a week before the planned union election, Jiron and her fellow workers were all called into a conference room. "They told us LCF was closing that day. For me, everything fell apart. . . . I will always carry around the fear of being fired and I will remember the threats to close if we voted for the union."[1]

Jiron, along with her 235 Spanish-speaking coworkers at Sprint's La Conexion Familiar (LCF) subsidiary in San Francisco, lost their jobs on July 14, 1994. On that day, Sprint shut down the telemarketing center before the union election that would have made this the first CWA bargaining unit of Sprint long-distance employees. Sprint was determined to keep the union out and used every tactic in the anti-union playbook to intimidate the largely female, immigrant workforce. When it appeared that a union victory was at hand, Sprint dealt its final card and shut the facility down. Sprint sent a chilling message to its workforce: a fight for a union will cost you your job.[2]

The National Labor Relations Board (NLRB) charged Sprint with illegal shutdown of the facility, and a Mexican union filed a complaint against Sprint under provisions of the North American Free Trade Agreement

(NAFTA). But in the end, a federal appeals court rejected the NLRB complaint, handing Sprint a victory and communicating to Sprint's unorganized employees that the nation's labor laws—even under a Democratic White House and sympathetic labor board—would do little to protect them. The Sprint workers' organizing campaign lost momentum. Sprint, which became a leading wireless carrier, was purchased by non-union T-Mobile in 2020, and it remained union free.

The Sprint long-distance unionization effort's failure stands in stark contrast to CWA's successful campaign during the same period assisting Southwestern Bell Mobile Systems (SBMS) employees to win union representation. Both campaigns took place among a predominantly female customer service workforce, many of whom worked in the South and Southwest. But at Southwestern Bell, CWA negotiated a "card check/neutrality" agreement with the company, allowing workers to organize outside the broken National Labor Relations Act (NLRA) electoral framework. Under the agreement, Southwestern Bell consented to remain neutral during any organizing campaign and to recognize the union on determination that a majority of the bargaining unit had signed union recognition cards. Absent employer intimidation and threat of job loss or retaliation, Southwestern Bell Mobility workers selected CWA representation, leading to the eventual organization of 45,000 wireless workers.

In this chapter, I analyze these two CWA strategic organizing campaigns to understand why CWA leverage proved effective in neutralizing employer anti-union animus at Southwestern Bell yet failed at Sprint. The differences in market competition and associated government regulation of Sprint's long-distance business and Southwestern Bell's local telephone market provide the primary explanation for both the level of intransigence in the employer's anti-union behavior and the widely different opportunities available to CWA to leverage political power to defuse the employer's resistance. In the years covered by these two organizing campaigns, Sprint operated without government oversight as a result of Federal Communications Commission (FCC) decisions to deregulate the long-distance telephone market and give favorable treatment to new entrants like Sprint. As a result, CWA had few opportunities to intervene in regulatory decisions that would affect Sprint's business. In contrast, state public service commissions continued to regulate Southwestern Bell's local telephone operation. Because this was a period of major regulatory reform, CWA had multiple opportunities to leverage its political power over state government decisions that would have major impact on Southwestern Bell's business success.

Moreover, Sprint and Southwestern Bell operated in distinct market structures, with different corporate cultures and levels of union strength.

Sprint remade itself in the 1990s from a corporation with dozens of rural local telephone subsidiaries into a long-distance company, competing against AT&T and MCI. Union representation at the local Sprint companies was divided between CWA and the International Brotherhood of Electrical Workers (IBEW), scattered across multiple states in small bargaining units, with little national presence or power. Sprint's long-distance call centers were greenfield operations with no union presence. Sprint management was determined to keep it that way to gain competitive advantage over unionized AT&T's higher labor costs and to maintain parity with non-union MCI.

By contrast, CWA had a strong union foothold at Southwestern Bell representing 37,000 non-management employees in the local telephone business, the major SBC revenue stream in the 1990s when mobile telephony was in its infancy. CWA had a mature fifty-year collective bargaining relationship inherited from the Bell system monopoly era and therefore directed its organizing at the employer, with a "bargain to organize" strategy that leveraged its power and membership strength on the telephone side of the corporation to negotiate card check/neutrality provisions covering the unrepresented wireless workers. This strategy proved successful at Southwestern Bell but failed at Sprint, forcing Sprint workers to seek union representation under the flawed NLRA election rules.[3]

In both cases, CWA and the organizing workers waged creative, multi-faceted campaigns that featured many of the characteristics labor scholars and frontline organizers have shown lead to union revitalization efforts.[4] But as these cases show, such campaigns may be necessary, but not always sufficient. In an environment in which employers can threaten workers' organizational efforts with plant closure and retaliation with virtual impunity, most workers choose a job over a union.[5] Call center employees are particularly vulnerable. With little capital invested in plant and equipment and a technology that allows work to be moved overnight anywhere across the globe simply by reprogramming the call distribution switch, call center employers possess an extraordinary ability to threaten job loss to block their employees' attempt to unionize. This proved fatal for the brave workers at Sprint's La Conexion Familiar.

The Organizing Imperative

CWA responded to the challenges of competition, deregulation, and technology change with what industrial relations scholars term a "revitalization" of the union, including an organizing program integrated with collective bargaining and political action. CWA's revitalization efforts centered on educating leaders and members to make organizing rights at the growing non-union Bell subsidiaries a top collective bargaining priority with the union-represented telephone companies. CWA developed a program of member mobilization to leverage power at the bargaining table and in the political arena to win employer neutrality in organizing to gain "wall to wall" representation at the growing non-union Bell company subsidiaries.[6]

Competition in the telecommunications industry came first to the long-distance market. By 1992, AT&T had lost about 40 percent share in the consumer and small business long-distance markets to MCI, Sprint, and others. AT&T responded by cutting 125,000 union-represented jobs, moving many technical and sales jobs out of the bargaining unit, and adopting a containment strategy that walled off acquisitions and new subsidiaries from union representation. Union density at AT&T plummeted from 62 percent in 1984 to 39 percent in 1992 and fell further to 25 percent in 1995.[7]

Competitive pressures in the 1990s were less intense in the local telephone market, which remained a regulated monopoly. Union density at the Regional Bell Operating Companies (RBOCs) stayed high at about 65 percent throughout the decade. Even so, RBOC management adopted cost-cutting and union containment strategies, as regulators rewarded "efficiencies" with the opportunity to earn higher profits, and competitors took market share in the lucrative metropolitan business market. Between 1984 and 1992, the RBOCs cut 158,000 union jobs at the local telephone companies and added about 75,000 non-union jobs in new wireless, data, and international lines of business.[8]

Unlike steel or auto in the latter part of the twentieth century, the telecommunications sector during this period was dynamic and growing. But as union density declined, CWA recognized the need to develop a strategic organizing program while the union still had majority representation in the industry, with the resources, political relationships, and membership strength to support large-scale organizing and set standards in negotiations. But as CWA and the customer service workers seeking union recognition discovered, their employers were determined to build their businesses union-free.

Sprint Campaign

In the early 1990s, U.S. Sprint Communications was the third largest U.S. long-distance carrier, with $9.5 billion revenue split evenly between the new long-distance division and rural local telephone companies. Sprint employed 43,000 people, of whom 16,000 (39 percent) worked for long distance, none with union representation. CWA and IBEW represented about 13,000 technicians in multiple bargaining units at Sprint's local telephone companies.[9]

CWA considered its union base at Sprint's local telephone companies a strategic resource and in 1990 launched a national campaign to assist Sprint long-distance workers to form a union. First, CWA leadership focused on educating local union leaders about the importance of organizing the non-represented side of Sprint. Many local union officers saw their primary role as representing the dues-paying members who elected them to office. CWA President Morton Bahr explained: "We have two choices. . . . [E]ither organize Sprint and bring their wages and working conditions up to our level, or face continued pressure to bring our wages and working conditions down to the level of Sprint." CWA organizers used a powerful metaphor. "[I]t is as if Sprint and other unorganized telephone workers are on management's side of the bargaining table—with their wages and working conditions pitted against ours. We are organizing Sprint to get these workers on *our* side of the table."[10]

CWA structured the Sprint campaign as a nationally coordinated network of local campaigns in dozens of cities across the country. Why national? "With the current technology in the industry, Sprint can easily move work from one location to another," the CWA organizers' memo explained. And why locally based? "Organizing is best done by Local Unions in their communities," the newsletter continued. This strategy reflected the philosophy of CWA's national organizing director, Larry Cohen. Cohen began his union career as director of the New Jersey state workers organizing campaign in the 1980s. There he built powerful worker-led committees (dubbed "The Committee of a Thousand"), leading to the eventual organization of 100,000 state and local New Jersey employees into CWA. In 1986, Bahr appointed Cohen national organizing director, tasked with the challenge of growing the union in the post-divestiture era. Cohen rose from this position to serve as an elected national officer, first as executive vice president in 1998, and then as union president from 2005 to 2015. As organizing director, Cohen developed a staff of highly committed, skilled regional organizing coordinators who were responsible for identifying, training, and supervising a network of CWA local member activists to support workers in their organizing campaigns. For Cohen, workers organizing workers was the key to the "transformational [experience] that is critical if working class people in this country are going to have a chance." Cohen summed up his philosophy: "Workers organize, unions support."[11]

CWA designed the campaign to "use the strength of our union—600,000 members in every town in America—to work on selected Sprint organizing targets coast to coast and develop ties and maintain long term contact with Sprint employees in those locations." The organizers' memo emphasized that "Sprint employees must be active in leading and recruiting others—we can't do it from the outside." When CWA had sufficient support within an organizing group, it would "go for recognition, either by election or neutrality card check (if that is obtainable)." CWA recognized that "Sprint management has a national anti-union campaign," and therefore the organizing effort required a long-term commitment with "CWA locals taking the lead . . . so that the network and Sprint union supporters have an organization that will survive whether or not a majority supports it to begin with." "The idea," Cohen later explained, "was to build the Sprint Employee Network, not rush into NLRB elections. It was clear that in 1991, that would be sending people to the slaughter."[12]

CWA organizers focused their outreach on the operator and customer service call centers, the two largest occupational groups. The call center environment posed unique challenges for organizers. Workers sat at their workstations, taking one call after another with little if any downtime between calls. Supervisors monitored their conversations with customers, and computers tracked the time spent on a call or offline. Opportunities for even a few minutes to socialize on the job were rare.

As CWA local activists reached out to Sprint employees, Sprint management accelerated its anti-union campaign. The corporation's "Union Free Management Guide" instructed supervisors that "US Sprint will face a myriad of challenges as we progress into the 1990s. Of these challenges, one of the most serious we face is the threat of union intervention in our business. . . . As management, you are expected to support US Sprint's union-free philosophies and programs. . . . There is no greater measure of your managerial effectiveness than a union organizing campaign." Tunja Gardner, a union activist employed as a customer service representative at Sprint's Dallas, Texas, sales center, described the impact of Sprint management's anti-union activities. "We have had numerous mandatory anti-union meetings, anti-CWA videos, memos from management and veiled threats. These activities are coordinated by our corporate Human Resources Department in Kansas City. We are bombarded with intimidating rumors. If an employee is fired, there's bound to be a rumor that she was fired for union organizing. There's always a rumor that Sprint will close our center to avoid the union. Everybody I work with has been made aware of the fact that the MCI closed its Detroit office a few years ago because the employees were organizing the union. . . . As a result, many of my co-workers are intimidated and afraid to participate. Let me be clear about what I mean by intimidated: they're afraid they'll lose their jobs or jeopardize their careers." The fear of job loss was not an idle threat. As Sprint automated and downsized its operator service operation, it closed centers with strong pro-union activists.[13]

Despite management's intimidation campaign, hundreds of Sprint call center employees joined the Sprint Employee Network, which officially launched in April 1993. In a widely circulated brochure, Sprint customer agents (Sprint's title for customer service employees) announced: "We're Sprint Customer Agents and We're Organizing a Union. Here's Why." The brochure featured pictures of fifty-two agents with statements explaining why each wanted a union. For some, the issue was fairness, the end to favoritism, and the need for clear procedures. "With CWA we can make sure there is an opportunity for everyone to advance based on their qualifications, not based on who's the bosses' favorite," said Cathy Berzinski of Winona, Minnesota. "I want CWA so we can't be pushed around anymore—so that we will have someone to turn to who will be on our side and stand up for us," added Karen Gellrich of Jacksonville, Florida. For others, the issue was the ability to join together to reduce stress, abusive monitoring, and inflexible attendance policies. "I like my job but I don't like my ulcers," wrote Mattie Jones of Nashville, Tennessee. For others, the issue was higher pay and job security. "I look at operators at the phone company and I know that we work as hard as they do . . . but we're paid so much less," noted Yvette Cotman of Richmond, Virginia. Sandy Johnson of Kansas City, Missouri, summed it up: "There are three good reasons to support the union—equal advancement, equal opportunity, and equal pay."[14]

Sprint activists were aware of the higher pay and benefits at the unionized AT&T and local Bell companies. Union-represented service representatives earned 33 percent more and operators 25 percent more than Sprint employees in comparable job titles. Moreover, Sprint used a complicated merit system to determine annual wage increases, whereas union employees' annual wage increase was contractually guaranteed. Sprint's higher health premiums and co-pays and inferior pension benefits widened the gap.[15]

By early 1994, Sprint Employee Network committees were pressing management for improvements in eleven centers. While the activists won some changes, they were not strong enough to gain majority support in any location except one: Sprint's La Conexion Familiar telemarketing center in San Francisco. It was here that the largely female, mostly immigrant workforce filed for an NLRB election, one that was poised to crack open the door to become the first unionized center in Sprint's long-distance business.[16]

In February 1994, several employees at Sprint's La Conexion Familiar (LCF, "the family connection") San Francisco call center walked across the

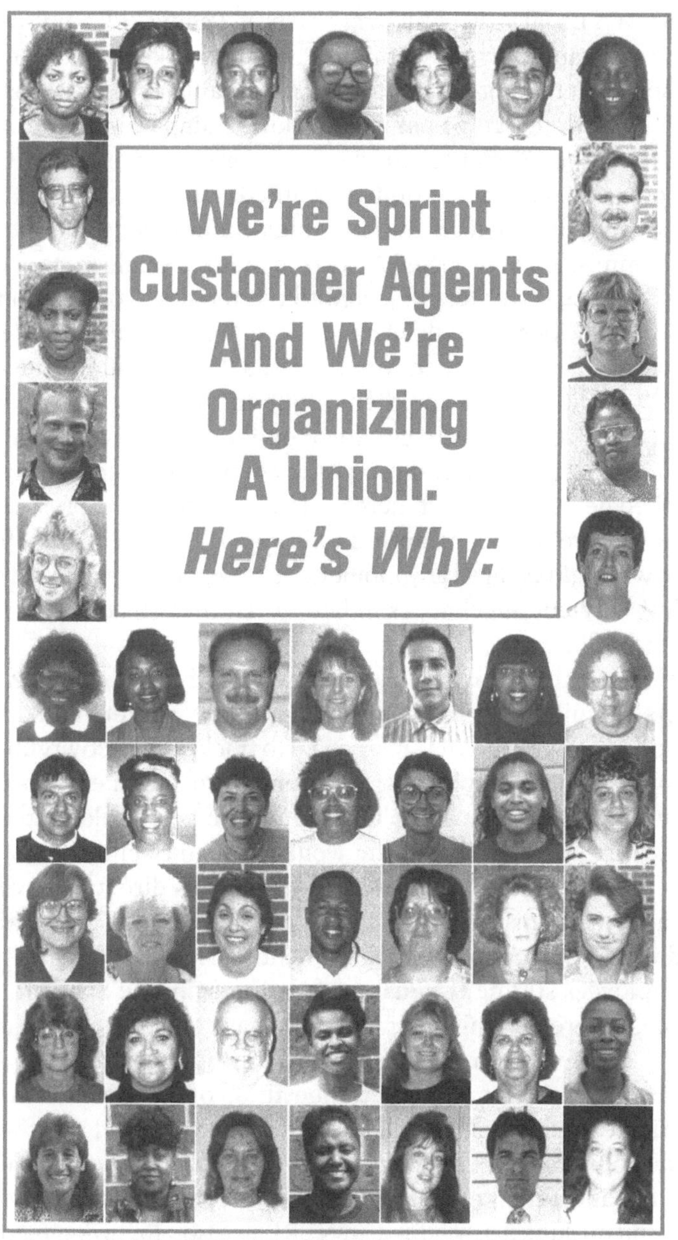

CWA pamphlet distributed as part of Sprint organizing campaign, CWA research department, circa 1992.

street to CWA Local 9410, which represented Pacific Bell employees, and asked for assistance forming a union. The LCF workers placed outbound calls to sell Sprint's long-distance service to Spanish-speaking customers. At $7 an hour, the largely Spanish-speaking immigrant workers earned $4 less than other Sprint telemarketers. Moreover, Sprint failed to pay them promised sales commissions. "I wanted a union because when I was hired they told me I'd get $7 an hour plus commissions and they didn't pay me commissions," reported Ana Hernandez, a single mother of three children. The LCF workers objected to the terrible working conditions and rigid work rules. "People were afraid to log off [the computer] to go [to the bathroom]," explained Myra Arriaga. "They told us to wait until the last possible moment and then to put our hand up, like we were all children. And they told us to cut down on fluids so we wouldn't need to go so much." Dora Vogel never received a commission because every time she reached her quota, Sprint increased the number. "The pressure to sell was great and we were constantly monitored. . . . [W]e felt stress from the fear that we might be fired any moment. . . . " Vogel was particularly upset by Sprint's last-minute changes in her work schedule to Saturday when she had no childcare.[17]

Within a month, Sprint began its aggressive anti-union campaign. Sprint managers recruited employees to spy on fellow workers and distribute anti-union material; required supervisors to track union activity and report names of pro-union employees; threatened pro-union employees with termination and discipline, including one firing; and told employees at six mandatory meetings that Sprint would close the facility if they elected union representation. The Sprint labor relations director flew in from headquarters in Kansas City, Missouri, to reinforce this message. Sprint later admitted that company management fabricated a flyer distributed to all employees stating (English translation from the Spanish): "many companies where the union has gone into have had to make the decision to close down operations and to move to other states or cities where no one is causing problems. You'd better be afraid of La Conexion moving to another state or city and we will be left without jobs and what will happen to our children? What do you need a union for? To die of hunger?" Despite the anti-union campaign, CWA was confident of strong worker support and on June 3, 1994, filed an NLRB petition for union recognition for a unit of 177 telemarketers and customer service representatives. By CWA's count, 70 percent of the workers favored the union. The NLRB set the union election for July 22, 1994.[18]

On July 6, 1994, two weeks before the scheduled vote, the Sprint board of directors voted to close the LCF facility, citing financial concerns. The reason for the LCF shutdown would become the subject of a three-year legal battle. One week later, on July 14, LCF management announced over a loudspeaker and in a handwritten letter that, effective that day, "your position will cease." All 177 telemarketers and customer service representatives lost their jobs. Within an hour, Sprint re-routed the LCF calls to Sprint's Customer Service Center in Dallas.[19]

CWA launched an innovative, multifaceted global campaign to support the Sprint workers. The CWA campaign was designed not only to win back the jobs and union election for the fired workers, but also to send a powerful message to Sprint and other anti-union employers that the threat of plant closings could no longer be part of their anti-union playbook. Most significant, CWA used the Sprint case to highlight to policymakers and progressive allies the need for fundamental reform of U.S. labor law.[20]

CWA filed an unfair labor practice charge against Sprint, asserting that the company illegally fired the LCF workers and closed the center to avoid a union election, violating Sections 8(a)(1) and 8(a)(3) of the National Labor Relations Act. The NLRB regional director agreed and filed a complaint against Sprint, seeking injunctive relief in U.S. District Court and an order to reopen LCF and reinstate the workers. The court denied the petition, but the NLRB complaint process moved forward.[21]

The NLRB legal process is notoriously slow, and CWA recognized that justice for the LCF workers required widespread mobilization of local, national, and even international political and community support. Sixty-four members of Congress wrote Sprint CEO William T. Esrey on the workers' behalf. Letters from women's organizations, Hispanic groups, religious leaders, labor unions, and state and local legislators flooded into Sprint headquarters. Sixty former LCF workers disrupted a San Francisco Board of Supervisors meeting to press for a hearing on the center closing. LCF workers sent an open letter to Sprint's spokesperson, actress Candice Bergen. (She never responded.) Vice President Al Gore met with a Sprint worker and President William J. Clinton called the Sprint labor-management situation "troubling." After Sprint inked a global alliance with Deutsche Telekom, German telecommunications union leaders raised concerns to Sprint officials.[22]

The most innovative tactic came from an alliance between CWA and the Mexican Telephone Workers' Union, STRM (Sindicato de Telefonistas de la Republica Mexicana). In February 1995, STRM filed a complaint against Sprint under provisions of the North American Free Trade Agreement (NAFTA). The complaint was grounded in the NAFTA labor side agreement, which gives any interested person in a signatory nation the right to file a complaint when another party to the treaty fails to enforce its own labor laws. In filing the complaint, the STRM/CWA partners turned the NAFTA labor agreement—initially designed to protect U.S. and Canadian workers from Mexico's notorious failure to enforce its own labor laws—on its head. The STRM complaint claimed that the slow judicial process under U.S. labor law and the failure to restore promptly LCF workers' rights demonstrated the ineffectiveness of U.S. labor law in violation of the NAFTA pact. The NAFTA labor agreement bars dispute resolution and limits remedial action to "ministerial consultation." Despite these weak provisions, STRM and CWA viewed the complaint as an opportunity to mobilize international pressure, on the Clinton administration to take domestic action against Sprint and on the Mexican government to condition Sprint's planned entry into the Mexican telecommunications market on respect for labor rights and reinstatement of the LCF workers.[23]

Ultimately, the STRM complaint revealed the weaknesses of the NAFTA labor side agreement. In December 1995, ten months after STRM filed its complaint, U.S. Secretary of Labor Robert Reich and his Mexican counterpart announced the outcome of their ministerial consultation: a tri-national investigation (along with Canada) into the effects of sudden plant closings on workers' freedom of association and right to organize. In addition, the U.S. Labor Department would host a forum on plant closings and the LCF case in San Francisco. In February 1996, one year after STRM filed the complaint and nineteen months after the LCF shutdown, the Department of Labor convened the forum. The hearing garnered significant publicity, as major U.S. news outlets reported on the irony of "U.S. Labor Making Use of Trade Accord it Fiercely Opposed."[24]

In June 1997, the Commission for Labor Cooperation issued the Plant Closings report. CWA blasted the document, which the union said "failed completely" to mention the Sprint LCF case or to "suggest and propose real reforms to protect workers' rights." To be fair, the Plant Closing report cited a commissioned study that found employers threatened workers with plant closing in half of the sampled union organizing campaigns and noted that employer threats of plant closing "can have adverse effects on workers' freedom of association and right to organize." Yet, the report offered no recommendations for reform. In the final analysis, the STRM complaint proved to be a paper tiger.[25]

The 1997 Plant Closing report came as CWA and the fired LCF workers were waiting for NLRB resolution of their case. Two years earlier, an NLRB administrative law judge (ALJ) found Sprint guilty of fifty labor law violations but failed to sustain the NLRB's charge that Sprint closed the center to avoid the union election. While recognizing that "LCF employees had been bombarded with statements by local LCF managers and supervisors that LCF . . . would be closed if the Union got in," the ALJ nonetheless concluded that Sprint closed the facility for financial reasons. During the trial, Sprint witnesses testified that LCF had a declining customer base, lost $4 million in the months before the closure, and projected an additional $7 million loss that year. The NLRB countered that Sprint continued to invest in LCF's turnaround in the months before the union election, hired new management and additional telemarketers, and refurbished the center. The most damning NLRB evidence was a forged letter—dated three months before the center's closing—from Sprint's vice president of human resources, seeking outplacement services for LCF workers in anticipation of a facility shutdown. The NLRB argued that the manufactured letter demonstrated Sprint's real reason for the center closure. The NLRB did not convince the ALJ on the plant closure charge, although the ALJ sustained the complaint's other labor law violations. His remedy, however, was toothless. He ordered Sprint to notify the fired employees that Sprint would no longer harass, threaten, interrogate, or spy on them—small comfort, since they no longer worked for Sprint.[26]

The NLRB, with CWA as an intervening party, appealed the ALJ decision. In December 1997, a three-member NLRB panel "amended" the ALJ decision to conclude that Sprint would not have closed the center "in the absence of union considerations." The NLRB ordered Sprint to rehire the fired workers and pay them back pay and benefits, with interest, which CWA calculated at more than $10 million. Sprint appealed the decision to the D.C Circuit Court of Appeals. The court, in November 1997, with liberal Judge Patricia Wald presiding, reversed the NLRB decision, finding that NLRB's "circumstantial evidence" was insufficient to counter the "overwhelming evidence that LCF was in a serious and sustained financial decline throughout the months before its closure." The fired Sprint workers were not entitled to any restitution.[27]

The court decision ended the CWA campaign for justice for the LCF workers and cast a dark shadow over CWA's efforts to support other Sprint long-distance workers in their union-organizing efforts. The campaign at Sprint long-distance waned as union activists found it increasingly difficult to build support among fellow workers who feared that union strength might make their center the next target for a shutdown. Over the next decades, CWA continued its corporate campaign against Sprint, successfully blocking its proposed $129 billion merger with MCI WorldCom in 2000, but the union was never able to break into Sprint's long-distance or wireless business. The challenges CWA faced at Sprint call centers, where an anti-union employer was able to block worker organizing campaigns with a flick of a switch, reinforced CWA's views that employer neutrality was essential to support workers who want a union. But as CWA's five-year campaign to win card check/neutrality at Southwestern Bell Mobile Systems demonstrates, a successful bargain to organize strategy took a unique combination of political leverage, union leadership and member mobilization, and top-level management support at a crucial time in the restructuring of the local telecommunications regulatory framework. These conditions cannot be replicated easily.

Five Years to Card Check at Southwestern Bell Mobile Systems

In the decade after divestiture, as AT&T and the RBOCs downsized their unionized telephone workforces and set up non-union subsidiaries for their growth businesses, CWA responded with a "wall to wall" program to pressure the companies to extend union recognition to their unrepresented workers. Central to the program was the bargain to organize strategy, in which CWA aimed to leverage its power on the unionized side of the corporation to negotiate provisions that would strengthen its ability to gain collective bargaining representation of non-union employees. The goal of the bargain to organize program was to return to the labor-management practices of the pre-divestiture Bell system, in which the union grew as the company grew, and optimally to establish a union recognition process outside the NLRA election framework.

The CWA bargain to organize strategy drew on earlier successful efforts by the United Auto Workers (UAW) and the International Ladies Garment Workers Union (ILGWU). In the 1970s, the UAW successfully pressured General Motors to grant union recognition under the "extension of operations" contractual language as the automaker opened new assembly plants in the South. The UAW strategy fell apart as the automaker divested and outsourced parts manufacturing, sent auto assembly offshore, and refused recognition in its joint ventures with foreign automakers. The ILGWU also used its bargaining strength in the North to gain recognition of garment employers and their contractors in the South. Stephen Lerner, lead organizer at the Service Employees International Union (SEIU), adapted the ILGWU strategy in the union's innovative Justice for Janitors campaign in organizing building service and security companies. UNITE/HERE has also used bargain to organize tactics to win union representation at unrepresented Hilton and Hyatt hotel chains.[28]

The success of CWA's bargain to organize program rested on four factors: first, a strong union presence in the corporation; second, a comprehensive union education program to mobilize local union leaders and activists to make organizing rights a bargaining priority; third, strategic opportunities for political leverage to pressure the company; and fourth, corporate management's attitude toward the union. These factors came together most successfully in CWA's campaign to organize wireless workers at SBC Communications, the parent to Southwestern Bell Telephone Company (SWBT), which provided local telephone service in Texas, Missouri, Oklahoma, Arkansas, and Kansas.[29]

In 1992, when CWA launched its wireless organizing campaign, CWA power at SBC was strong, representing 37,000 telephone employees, 63 percent of the SBC workforce. SBC was a profitable company, and the local telephone business composed 80 percent of revenue. CWA had multiple opportunities for political action, as policymakers considered market-opening telecommunications reform and regulators reviewed controversial SBC mergers. CWA's regional vice president, Vic Crawley, made wireless organizing a top priority after attending a shareholders' meeting in which the CEO reported that SBC's strategic plan projected that mobility would compose 70 percent of SBC revenues in ten years. Finally, SBC's chief executive, Edward Whitacre, who started his career at Southwestern Bell Telephone as a union-represented technician, recognized the value of a labor-management partnership and intervened at crucial moments.[30]

Initially, SBC wireless management responded to CWA's wireless organizing with an anti-union campaign, hiring consultants to guide local managers in union-avoidance tactics; transferring, firing, or demoting union activists; packing bargaining units with anti-union new hires; highlighting weaknesses in existing union contracts; characterizing the union as an outside third party; and creating an uncomfortable atmosphere of conflict in the workplace. Under these conditions, Southwestern Bell wireless workers were reluctant to vote for union representation. CWA had a 3–3 NLRB win-loss record in the first five years of the campaign. But ultimately management's anti-union strategy backfired. Union activists were deeply offended that "Ma Bell," the corporation that had given them and often members of their family good jobs, would actively oppose the union. CWA members' and leaders' resolve solidified around a multi-pronged strategy of member education, worker mobilization, political action, and negotiations to win neutrality and card check recognition at SBC's non-union subsidiaries. Under card check recognition, an employer agrees to recognize the union upon presentation and third-party certification of union authorization cards signed by a predetermined number of workers (typically 50 percent plus one). Under neutrality, the employer agrees not to "help or hinder" the decision of a group of workers on union representation.[31]

In 1992, fresh from the SBC shareholder meeting, Crawley launched a district-wide campaign to organize Southwestern Bell Mobile Systems (SBMS). Two seasoned organizers, Danny Fetonte and Sandy Rusher, led the effort. Fetonte and Rusher honed their organizing skills building the CWA-affiliated Texas State Employees Union (TSEU), a public-sector union in a state without public employee collective bargaining rights. In building TSEU, Fetonte and Rusher learned many lessons that would prove crucial at Southwestern Bell Mobile Systems: the need to build strong worker-led organizing committees; the importance of building alliances between CWA's organized telephone workers and unorganized workers; and the critical role that political action could play in leveraging power. Using the CWA worker-led organizing model, they began to train a network of local union activists to reach out to Mobile Systems workers.

CWA Vice President Crawley took the lead educating union leaders and activists about the wireless organizing imperative. "I don't think there was a meeting I was part of where I didn't include a discussion of the direction our industry was heading and why it was critical for us to organize wireless," Crawley later recalled. Crawley's "Why Organize?" presentation used powerful graphs to show union job loss at Southwestern Bell Telephone Company at the same time that non-union jobs were growing at SBC's subsidiaries. Crawley hammered home the message that CWA would not be able to provide good representation to its members if the union did not bring unorganized wireless employees into the union. In the first two years alone, 350 member activists attended educational workshops and fifteen local unions got involved in wireless organizing.[32]

The campaign launched in St. Louis in 1992 with a meeting between CWA activists and Mobile Systems employees inside a building with both telephone and wireless employees. CWA local organizers moved relatively freely throughout the building, talking to workers and distributing literature. The union activists learned that 90 percent of wireless employees work in customer service, evenly split between call centers and retail stores. (Wireless technology does not require the large construction and maintenance workforce of a landline network.) The union workers also discovered that SBMS customer service representatives earned less than half the wages of their unionized telephone company counterparts, and also had higher contributions for health benefits. These facts sharpened their resolve. "I had worked for SBC all my life and it was a shock to hear about the wages and working conditions of the mobile workers," Tena Ryland, CWA local president in Abilene, Texas, explained. "This was the fastest growing entity in the company and I knew that as they got larger they would weaken us. I knew that by organizing these workers we would be able to improve their conditions and also protect our own in SBC."[33]

St. Louis Mobile Systems management responded to the CWA outreach with anti-union information. In response, Crawley, who was negotiating the 1992 telephone company contract, pushed hard at the bargaining table to win a corporate-wide neutrality agreement. He won half a loaf: SBC agreed to neutrality covering all SBC telecommunications subsidiaries *except* Mobile Systems. Later that year, CWA and Mobile Systems CEO John Stupka signed a letter with a neutrality clause covering Mobile Systems, giving the union access to wireless employees on company premises. Mobile Systems managers violated the agreement. After CWA filed for an NLRB election in St. Louis in 1993, Mobile Systems announced mandatory employee meetings to deliver its anti-union message. CWA sprang into action. Crawley reached out to CEO Whitacre, and the CWA Texas legislative director told SBC that CWA would withhold support on pending telecommunications legislation if anti-union meetings continued. Although Mobile Systems did not cancel the mandatory meetings, SBMS management allowed a CWA staff representative to make a presentation about the union at the meetings. This proved critical. "I think we would have lost the election at that point if we had not been able to show the mobile workers that we had some power through our relationship with the company," Crawley recalled. CWA won its first SBMS election in St. Louis, 59–34.[34]

The next stop was Dallas/Fort Worth, where Southwestern Bell Mobile Systems had twelve locations and about 650 employees. After a twenty-five-person internal organizing committee posted a pro-union letter on company bulletin boards, SBMS CEO Stupka hired an anti-union consultant to train local supervisors and transferred or promoted union leaders out of the bargaining unit, effectively cooling the organizing effort. CWA organizers shared the consultants' anti-union handbooks and activities with CWA local leaders. This angered CWA leaders, who were accustomed to a telephone company management that largely accepted the union, and it deepened their commitment to win organizing neutrality.[35]

CWA activists, organizers, and staff formed a district-wide strategy group that frequently debated—and constantly re-evaluated—whether to mobilize political support or opposition to SBC public policy goals. As SBMS increased its union avoidance activity, local leaders became more willing to oppose the company in the political arena as state legislators and regulatory commissions debated telecommunications reform. In 1993, CWA leaders in Kansas supported SBC in the legislature, but only after SBC signed a no-layoff pledge for union members. But a year later, the president of the powerful Dallas local urged fellow Texas local union presidents to withhold support for SBC-supported telecommunications legislation, reasoning that if SBC blocked union growth in wireless, the union should block SBC's growth in the state. CWA staged a large demonstration at San Antonio corporate headquarters protesting Mobile Systems anti-union activity.

Passage of the market-opening federal Telecommunications Act of 1996 intensified the state regulatory battles. To jump-start competition, the 1996 act required state regulatory commissions to set the wholesale rates that incumbent telephone carriers like SBC could charge competitors to lease all or parts of the SBC network for resale. SBC reached out to CWA to line up support. CWA district and local leaders told SBC management that the union would not support policies to help grow the company as long as SBC blocked union growth in wireless and other non-union subsidiaries. "We had a retreat with SBC to talk about our relationship with them, and a large amount of time was spent on the question of Mobile Systems, why we believed we should have recognition but would accept card check," reported Alma Diemer, president of a Little Rock, Arkansas, local. The message got back to the top levels of the company that CWA would not support SBC's position on Arkansas telecommunications legislation without company movement on card check. In other instances, CWA demonstrated its political clout and value to SBC by supporting telecommunications legislation that would benefit SBC and the workforce. In 1995, CWA brought 5,000 members to Austin, Texas, to lobby legislators on a major telecommunications reform bill. Before lobby day, CWA local leaders told SBC management that the union was looking for card check and recognition at wireless. Elsewhere, the San Antonio local convinced the city council to drop a plan to lease a city-owned fiber ring to an SBC competitor. CWA staff Gloria Parra would later remind SBC CEO Whitacre that such support would not come again if SBC continued to fight the union at Mobile Systems.[36]

As CWA continued reaching out to wireless workers in Texas, SBMS management consistently violated the neutrality agreement with mandatory anti-union meetings and dismissal of pro-union activists. After CWA lost an NLRB election in Abilene, Texas, in 1994 by two votes, the union filed unfair labor practice (ULP) charges. Union organizers used the NLRB hearings to gather information about the company's anti-union activity, sharing details with activists, fueling their anger and cementing support for neutrality/card check at wireless. The continuous pressure began to pay off. To get the Abilene ULP charges dismissed, SBC agreed to its first card check agreement, giving recognition in Abilene if 60 percent plus one signed representation cards certified by a neutral third party. With card check, CWA signed up 70 percent of the Abilene unit over eighteen months. CWA continued to press SBC and Mobile Systems executives to strengthen and abide by the neutrality agreement, and in 1995, the parties inserted a mediation and arbitration provision into a new neutrality agreement. This made a difference. In May 1995, Mobile Systems management refrained from anti-union activity before an NLRB election in Corpus Christi, Texas, and stopped an employee from passing out anti-union material on the job. The union won the election.[37]

Soon after, CWA and SBC signed a 55 percent plus one card check/neutrality agreement covering Mobile Systems in Houston, Austin, and Beaumont, Texas. In exchange, CWA agreed to allow wireless employees to sell wireline services (work over which the union had jurisdiction) in one-stop retail centers. Seventy-five percent of Mobile Systems workers in Houston, Austin, and Beaumont signed up and were certified for CWA representation. Yet, Mobile Systems did not abide by the neutrality agreement in other locations. In San Antonio, the anti-union consultant trained managers to spread negative information about CWA and successfully thwarted the election, with CWA losing by twenty-five votes.[38]

CWA saw a golden opportunity to win system-wide card check/neutrality language when SBC announced in the spring of 1996 its merger with Pacific Telesis, the RBOC in California and Nevada. The passage of the market-opening Telecommunications Act of 1996 ushered in a wave of corporate restructuring, beginning with the SBC/Pacific Telesis merger. These transactions were subject to Federal Communications Commission (FCC) and state commission review. SBC had a high hurdle to surmount to convince regulators, especially the tough California Public Utilities Commission, to approve putting two former Bell system companies back together. CWA District 6 Vice President Ben Turn, who had replaced Crawley, and Vice President Tony Bixler in CWA District 9 (covering California and Nevada) agreed they would withhold merger support until their respective employers agreed to card check/neutrality in wireless. After many months of negotiation, in spring 1997 CWA Districts 6 and 9 signed comprehensive card check/neutrality agreements with SBC. The language covered all SBC lines of business and future in-region subsidiaries; required 50 percent plus one signed cards for certification; and gave automatic recognition to any unit where more than 50 percent of the workers were transferred from an already-represented CWA group. CWA then supported the SBC/Pacific Telesis merger on the basis that the extension of organizing rights throughout the corporation would give workers a collective bargaining platform to negotiate a fair share in the benefits of the combined company's growth.[39]

At its 1997 convention, CWA recognized District 6 with the coveted president's organizing award. CWA President Morton Bahr acknowledged "the enormous impact that this agreement can have on us and our future. . . . The pact . . . covers future Internet and video services, mobile systems or wireless systems, one-stop shopping stores, and things that are not even on the board yet." Bahr emphasized that "even under the extremely difficult conditions for organizing in the U.S., District 6 locals realized that wireless is the key future technology in voice communications and that organizing these workers is critical to the future of the movement. . . . [T]he entire district, led by two vice presidents, at least 100 local officers, and thousands of members has demonstrated that we do make a difference." Then President Bahr took an unprecedented step of inviting SBC CEO Edward Whitacre to address the delegates. Bahr praised Whitacre as a corporate executive who "sees CWA as adding value to the company" who understands that "good relations with their unions . . . and their employees were their most valuable asset."[40]

As SBC acquired other RBOCs over the next decade (Pacific Bell, 1997; SNET, 1998; Ameritech, 1999; AT&T, 2005; BellSouth, 2007), CWA conditioned support for these mergers on extension of the SBC card check/neutrality language to the newly purchased entities, a condition that SBC (renamed AT&T after the 2005 merger) accepted in each acquisition. Under card check/neutrality, SBC wireless employees signed up for the union, and as the wireless industry grew, so did the number of CWA-represented wireless employees, reaching about 45,000 AT&T Mobility wireless workers at its height in 2015, representing about one-third of the U.S. wireless workforce. But as AT&T accelerated outsourcing call center and retail store jobs, the number of CWA-represented wireless workers dropped to 32,000 in 2021.[41]

In both the Sprint and Southwestern Bell wireless organizing campaigns, workers and the union confronted substantial employer resistance. Ultimately, the wireless workers won their union, but the Sprint workers did not. At SBC, the wireless organizing took place at a critical moment of telecommunications regulatory reform and corporate restructuring, in a period in which the union's stronghold, the local telephone market, was just emerging from regulated monopoly to one of competition. This gave CWA multiple opportunities to mobilize its considerable political power in the states to win employer concessions to neutrality and card check recognition. Without fear of job loss or retaliation, Southwestern Bell wireless workers chose union recognition. But few, if any, of these institutional and market conditions existed at Sprint, which operated in a highly competitive, deregulated long-distance market, where political levers,

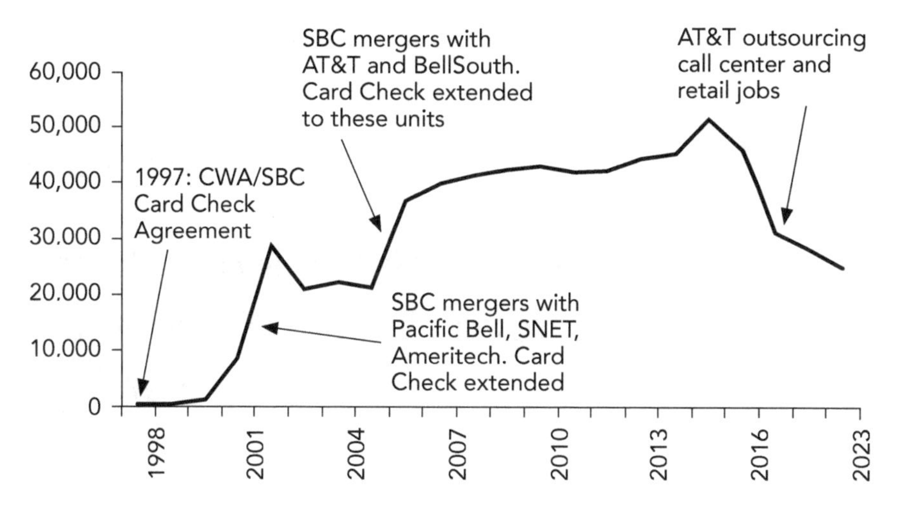

CWA-Represented Workers at SBC/AT&T Mobility, 1998–2023.

including the federal labor regime and the labor provisions in NAFTA, did not protect the Sprint workers who wanted a union. In the end, the weak NLRA framework allowed the employer to give workers an untenable choice: the union or your job. While the essential difference in these two cases boils down to institutional factors—the regulatory regime and degree of market competition that the two different employers faced in the mid-1990s—timing, geography, corporate culture, and company leadership were also critical.

The failure of U.S. labor law to protect non-union telecommunications workers seeking union representation marked the end to any hopes of reviving the industry-wide sectoral bargaining of the monopoly era. With the 1992 election of William J. Clinton as the first Democratic president in twelve years and with Democratic majorities in Congress, the labor movement pressed his administration and a Democratic Congress to pass fundamental labor law reform. But the Clinton administration never got behind the effort, choosing instead to focus on labor-management partnerships to give workers greater voice on the job. For a brief moment in the 1990s, AT&T and Bell Atlantic joined the labor-management bandwagon as managers struggled to redesign the top-heavy, bureaucratic work systems inherited from the Bell system. CWA leaders and call center activists saw opportunity to engage with their employers to redesign the service representative job, opening the door to what the union leaders considered an expansion of collective bargaining over fundamental issues of work organization and technology.

CHAPTER 4

False Promises? Job Redesign through Union-Management Partnerships

"Job redesign was a very meaningful time of my career," Linda Mulligan recalled as she described the eighteen months she cochaired the CWA/US West union-management project to revamp the service representative job. "The process created so much energy. The service representatives felt they were appreciated. Work pressures went down when the company did away with sales objectives. We came up with a job redesign that streamlined the process for the customer, that used the service representative's knowledge to make sure the customer got what was needed. Job redesign also built a stronger relationship between members and the union. Members saw the union involved in so many areas of problem solving. But the company never implemented the job redesign that we worked so hard on. I was left with feelings of disappointment, distrust, heartbroken, and betrayed."[1]

In this chapter, I analyze the effectiveness of yet another strategy that CWA and its customer service members pursued in the mid-1990s to improve working conditions in the call centers: union-management work transformation teams. I highlight initiatives at AT&T and two RBOCs—Bell Atlantic and US West—in which CWA leaders and top-level managers worked diligently to redesign call center operations. It was an exciting time for the union and its members. In each case, committed directors of the customer care operations, union leaders, and frontline workers developed models for call center operations that met management goals for more streamlined, customer-focused operations and union objectives of employment security, good compensation, and more varied, autonomous work. The moment was ripe for change as AT&T and Bell managers aimed to restructure their work organizations from the bureaucratic, management-heavy structures of the monopoly era to systems that were leaner, more flexible, and more customer focused for the competitive era.[2]

Labor-management joint initiatives are a controversial subject among union activists and students of labor relations. Critics argue that failure is baked into any joint labor-management work transformation program. Victor Reuther, who along with his brother Walter was one of the early founders of the UAW, slammed such partnerships in the auto industry as "an enticing illusion that the worker will have a real voice in management." These joint programs, he argued, served to "undermine the unity and solidarity of the workers on the plant floor." *Labor Notes* editors Mike Parker and Jane Slaughter authored an influential critique of the team concept in auto plants as a form of speed-up and "management by stress." Labor and management have adversarial interests, these critics assert; labor-management programs are simply a mechanism to give employers access to workers' knowledge to squeeze labor, cut jobs, and boost shareholder return at workers' expense.[3]

In this chapter, I reach a more nuanced conclusion. The CWA leaders who bargained and promoted labor-management initiatives in the telecommunications industry saw these joint programs not as cooperation or co-optation, but as collective bargaining on different terrain. Having suffered through a decade of company-driven job cuts in response to competition, technology change, and regulatory reform, CWA President Morton Bahr believed that the union had to push for a seat at the table at all levels of the corporation. "Over the years," Bahr explained at the 1993 kick-off conference for the negotiated joint Workplace of the Future program at AT&T, "CWA has been enormously successful in negotiating a range of options . . . to ease the impact on [our members] if their jobs were impacted by technology or by restructuring. . . . Now, with Workplace of the Future, the union is guaranteed a significant voice in managing change in the workplace . . . before decisions are made."[4]

This was the promise. The call center activists who poured precious time, energy, and their credibility as union leaders into union-management job redesign projects went into these programs with their eyes wide open. They saw this as an opportunity to achieve their own goals to enhance job security, improve working conditions, maintain a relatively high-wage compensation structure, and increase union presence and power in the workplace and at higher levels of the company. Leaders of the call center bargaining units understood that the dehumanizing and stressful conditions in the call centers were grounded in the work organization and associated electronic management systems that controlled the pace and manner of work. These union leaders saw union-management participation programs—if properly structured to give the union a real role in the process—as yet another arrow in their quiver, together with collective bargaining, contract enforcement, organizing, and member mobilization. They saw the union-management work teams as an opportunity to strengthen the union and its relationship with members, using company-funded meetings to bolster a network of union activists across widely dispersed regions, taking advantage of company-sanctioned time off the job to meet with and engage members, and deploying company-funded leadership training to enhance their own skills as union leaders.[5]

Certainly, some CWA members and leaders fell into the "illusion of inclusion" trap. But many did not. For the most part, the union call center leaders were not co-opted into acquiescence by these joint programs. As we shall see, they remained true to union goals throughout the team initiatives' duration. When corporate management ultimately rejected many (though not all) of their proposals in favor of job cutting, reengineering, and union avoidance strategies, CWA call center leaders and members did not abandon adversarial mobilization activities, particularly around contract fights. In 1995, the service representatives at Bell Atlantic joined their brothers and sisters in a five-and-a-half-month work-to-rule campaign, and in 1998 took strike action to win a good contract. In 2000, they resurfaced many of the demonstration programs piloted during the joint team process as bargaining demands in their strike against Verizon. Similarly, the CWA leaders and activists who led joint job redesign committees at US West went out on strike in 1998. And at AT&T, as we shall see, CWA used the union-management job transformation process to achieve a decades-long goal to end a second-tier wage rate in the business customer care centers.

Like collective bargaining, union-management committees can be arenas for the contest between labor and management. Like collective bargaining, the outcome of this contest depends on the relative power that the union brings to these programs. The experience of the union call center leaders and activists in using the joint committee process at AT&T and Bell Atlantic to improve conditions for their members helps us understand the limitations of such programs within the U.S. context. Unlike companies in Scandinavia, Germany, and Austria, U.S. firms are not legally required to engage with unions or elected works councils over key issues such as technology deployment and work organization.[6] The CWA leaders and activists who participated in joint work redesign teams pushed the limits of the voluntaristic structures, and in the process, discovered the boundaries of their power in this arena of contested terrain.

The New Mantra: High-Performance Work Systems

The AT&T and RBOC labor-management initiatives that I discuss in this chapter were part of a short-lived moment in U.S. capitalism and management practices in which a broad constituency of managers, union leaders, academics, policymakers, and some corporate leaders advocated for and experimented with employee participation and joint labor-management programs to design new work systems. In response to the economic challenges that began in the early 1970s, characterized by declining productivity, lower profit rates, and stagnating wages, as well as by worker and union desire for more meaningful, secure work, a significant number of companies, workers, and their unions promoted new work systems that engaged frontline workers in productivity and process improvements. Advocates of these new work systems argued that the assembly-line mode of production (and service delivery) based on high-volume manufacturing (and service delivery) of standardized products no longer offered U.S. firms the comparative advantage they had achieved throughout much of the twentieth century. Taking lessons from successful programs in Japan, Germany, and Sweden, these proponents argued that American competitiveness in a knowledge-based, global economy required replacing the dominant mass-production model—a work system that separated thinking from doing and structured work through detailed, repetitive division of labor—with organizations that gave frontline workers greater autonomy and input into the work process. Employers who were particularly impressed with Japanese quality circles and German flexible production believed that these new work systems would lead to a competitive model based on quality and customization rather than low-cost mass production.[7]

The AFL-CIO Committee on the Evolution of Work, which included the presidents of twenty-eight unions from the industrial, service, and public sectors, including CWA President Morton Bahr, endorsed labor-management partnerships. In its 1994 policy statement, entitled "The New American Workplace: A Labor Perspective," the committee encouraged unions "to take the initiative stimulating, sustaining, and institutionalizing a new system of work organization based upon full and equal labor-management partnerships." The labor federation saw new forms of work organization as an opportunity to replace demeaning Taylorism on the assembly line with new models of work that would reject the separation of thinking from doing; offer workers greater variety of work and responsibility; give unions decision-making roles at all levels of the enterprise;

and distribute rewards from enhanced productivity on equitable terms through collective bargaining. The AFL-CIO made clear that true labor-management partnerships must be based on the employer's respect for the union. These initiatives, if properly structured, could allow unions to overcome traditional contractual limits of "management rights" clauses "to insist upon the right of workers to participate in shaping the work system under which they labor and to participate in the decisions that affect their working lives."[8]

The Democratic Clinton administration embraced the vision of a new American workplace as the basis for a competitive, high-wage, high-skill economy, one that would reverse the negative impact of deindustrialization and foreign competition on U.S. workers and the economy.[9] Within the Clinton administration, Labor Secretary Robert Reich became the leading proponent of work transformation as the path to competitiveness and good U.S. jobs. "The only way you become competitive as a company in this new global economy . . . is to change your organization," he told the audience of 1,000 AT&T managers and CWA union activists at the Workplace of the Future opening conference. "The high value organization is an organization in which workers and managers are working so closely together, sometimes, it's impossible to tell the difference."[10]

For the Clinton administration, a program encouraging new work systems and labor-management participation would not only address U.S. competitiveness challenges but also give the administration a path to neutralize business opposition to labor law reforms to strengthen union organizing efforts. To this end, labor secretary Reich and commerce secretary Ron Brown appointed a thirteen-member Commission on the Future of Worker-Management Relations (known as the Dunlop Commission after its well-respected chair, former labor secretary and longtime labor-management specialist John T. Dunlop).[11] Dunlop aimed to craft a grand bargain between business and labor, pairing management's desire to amend the National Labor Relations Act to permit management-appointed employee committees in non-union settings with labor-supported changes to the NLRA that would make it easier for non-union workers to organize. Union leaders vehemently opposed any provision that would open the door to company unions as an undemocratic violation of the principle of independent worker organization. Even more significant, employers plowed ahead with management-dominated committees with no legal interference. They saw no need to strike a "grand bargain" that would empower unions by strengthening existing labor law. Dunlop's quid pro quo died an early death.[12]

CWA Evolving Views on Labor-Management Partnership

In 1980, CWA became an early participant in the first generation of labor-management participation programs when the union negotiated a Quality of Work Life program (QWL) in the AT&T contract.[13] The QWL program was designed "to improve the work life of the employees and enhance the effectiveness of the organization." The union saw QWL as a means to reduce the kind of job pressures, excessive monitoring, heavy supervision, and unreasonable output standards imposed on its members, especially the operator workforce. At the same time, AT&T was concerned with declining employee morale, evidenced by a 1980 survey that showed 40 percent of employees were dissatisfied with their jobs because "jobs were too structured, there was an over-abundance of measurement and too great an emphasis on employee productivity." With the assistance of psychotherapist-turned-organizational consultant Michael Maccoby, the union and company created the QWL program to humanize and streamline work processes. A joint union-management governing committee established principles for the QWL program: all workplace initiatives must be voluntary, the union must be an equal partner, QWL must not lead to layoffs or job downgrades, and the program must stay away from collective bargaining and grievance issues. AT&T committed $5 million for training and facilitators. By 1983, there were more than 1,200 problem-solving teams in place, hundreds of trained facilitators from management and employee ranks, and an estimated 100,000 employees who had participated in the program.[14]

A 1985 Department of Labor evaluation concluded that most improvements were "modest" and limited to cosmetic issues (redesign of the break room, for example).[15] In New York City, the large AT&T local pulled out of the program after the company closed the international operating center, displacing 2,000 operators. Local President Laura Unger concluded that "QWL is a process, which, under the guise of empowerment, trains workers to internalize the goals of management . . . often against their own self-interest as workers. . . . QWL actually takes away power from the workers."[16] There were some significant exceptions. In Tempe, Arizona, for example, the union and a progressive manager created an entirely worker self-managed operation in a brand-new hotel billing and information system office, leading to 35 percent supervisory cost savings, 45 percent profit margins, high worker morale, and decline in absenteeism and grievances. But AT&T shut down the center after divestiture, foreshadowing the corporate decisions to shut down the call center job redesign projects that I describe in this chapter.[17]

In 1983, CWA, AT&T, and the RBOCs expanded on the QWL program by negotiating "common interest forums" to bring together top union and company leaders to discuss high-level issues of common concern. Together with the technology change committees negotiated three years earlier, CWA leaders hoped these joint programs would lead to a larger role for the union in implementing new technologies and work organization. But they were short lived. The massive job cuts, dislocation, and turbulence that accompanied the breakup of the Bell system the next year effectively killed the spirit of trust and mutual gain that are essential to any successful labor-management program. As CWA President Bahr explained, "[T]he 1984 divestiture delayed the development of effective committees and unleashed a ten-year period of turmoil and disruption." With the notable exception of BellSouth and Mountain Bell (the latter is one of the operating companies that merged to form US West after divestiture), these workplace programs effectively died at AT&T, Bell Atlantic, and the other RBOCs.[18]

Yet, the lessons CWA learned from the QWL experience would guide the union in the 1990s as it entered into several negotiated labor-management programs designed to address the union's primary challenges in the post-divestiture period to ensure employment security for its members and to strengthen the union as an independent institution of worker power. The 1994 CWA executive board policy on union-management participation in the telecommunications industry made clear that the union wanted a seat at the table at all levels of the company, with access to vital information and an opportunity to influence key financial, technological, and human resources decisions during planning stages, rather than after the fact. "With effective union-management participation," the policy statement explained, "there will be more opportunity to protect workers at companies which are now being forced to change because of technology, further deregulation and increased competition." Union-management participation, the CWA statement concluded, "has become essential to effective collective bargaining in the information age."[19]

Workplace of the Future at AT&T

In 1992, CWA and AT&T reached agreement on an ambitious union-management program called Workplace of the Future, which became one of the most widely cited labor-management programs of the period. Workplace of the Future (WPOF) differed from the earlier Quality of Work Life program in several ways. First, the WPOF Agreement provided for union involvement at multiple levels of the company, including the workplace, the business unit (an organizational subdivision of the company), and the corporation. Union leaders and staff appointed all employee members of workplace teams and business unit planning councils, and all WPOF training and materials were jointly developed and implemented by the union and the company. Union goals for employment security were incorporated into the WPOF Agreement, recognizing that WPOF initiatives to target customer satisfaction and market flexibility must also "be sensitive to employees' needs regarding employment security." The WPOF contractual agreement stated that "[i]t is not the Company's intention for employees to be negatively impacted by workplace innovations resulting from employees' ideas."[20]

CWA entered into WPOF with one overriding goal: to save jobs. In the decade that followed the 1984 Bell system breakup, the union was unable to stanch the company's downsizing of 125,000 union jobs, despite a twenty-five-day strike in 1986. In preparation for 1992 contract negotiations with AT&T, as CWA President Morton Bahr put it, the union wanted to move beyond negotiations over programs that would "ease the pain" of AT&T downsizing and restructuring to gain a "guaranteed (and) significant voice in managing change in the workplace." A CWA survey conducted in preparation for 1992 bargaining found that 85 percent of members wanted the opportunity to have a voice in decision-making about their jobs.[21]

At the same time, AT&T managers were desperate to find a way to make the company more productive and customer focused, to reduce the high costs associated with its bureaucratic structure and large managerial workforce, and to gain greater flexibility in responding to customer demand. While AT&T's financial position had largely stabilized by the early 1990s, its cost structure remained high compared to its more nimble non-union competitors. Although some senior managers advocated decertifying the union, they became convinced that the cost in time, money, and disruption would be too high. Therefore, before 1992 contract negotiations, AT&T's vice president for labor relations, Bill Ketchum, turned to consultant Michael Maccoby to reach out to CWA Vice President Jim Irvine (who was responsible for the AT&T bargaining unit) to craft a more cooperative labor-management approach. The result was Workplace of the Future.[22]

CWA and AT&T launched Workplace of the Future with great fanfare at the 1993 conference. In addition to speeches by Secretary of Labor Robert Reich and CWA leaders Morton Bahr, AT&T Vice President for Data Communications Services Stan Kabala voiced his strong support for the program because "the key to our survival and prosperity lies in a revolution that's led by involved, empowered, and enthusiastic employees." For

Kabala, who was responsible for AT&T's data network, competitive success required overturning the Bell system's bureaucratic and costly work organization, with its multiple layers of management ensuring compliance with standard operating procedures. Pushing responsibility down to frontline workers could save costs not only by reducing management layers but also by capturing workers' knowledge to improve processes.[23]

CWA Vice President Irvine acknowledged that despite union "naysayers," the program went well beyond QWL, and he expressed optimism that the union and management could make this "revolutionary concept" work. Irvine added that nearly 90 percent of the membership ratified the 1992 contract, indicating rank-and-file support for the program. Laura Unger, militant president of Local 1105 in New York City, recalls that many local leaders were skeptical of WPOF. They recalled the failure of QWL to stop AT&T layoffs and rejected the lofty rhetoric that claimed workers and managers shared common ground. Despite her skepticism, Unger was willing to try it. "[O]ur members can't go on living in constant fear of losing their jobs," she told managers at an early WPOF meeting, and "management . . . cannot survive in a customer focused competitive environment with your employees and their Union constantly aiming for your jugular." Unger warned that WPOF could promote "an illusion of inclusion" unless the union developed a strong independent agenda and used the WPOF structure "as a tool to fight for our members."[24] Despite the skepticism among many local leaders and activists, there was no organized opposition to WPOF. Those who didn't support the program just did not participate.[25]

CWA national leadership and staff and many local leaders worked hard to make the program a success and to ensure that union goals were incorporated into Workplace of the Future initiatives. The national union provided local leaders with a guide to union involvement in WPOF. Thousands of union members and company managers attended WPOF training sessions, with a curriculum developed by Rutgers University School of Industrial Relations that included sections on union and CWA history as well as AT&T challenges in the telecommunications industry. For many local union leaders, WPOF demanded an enormous amount of time and oversight to ensure that union goals were not subsumed by management in the joint committees. Unger explained to a labor-management WPOF forum that she was working harder than ever doing her traditional local president tasks with the added load of WPOF; sitting on a planning council, the business communications service (BCS) customer service transformation effort (described below), two planning council subcommittees, and one cross-planning council subcommittee; reading the minutes of every WPOF committee in her local; and giving training classes to management.

The Workplace of the Future program lasted seven years. The results were mixed. "I don't recall it achieving anything," Colleen Dowling, president of the large Lee's Summit, Missouri, call center, later recalled. "Our attendance committee tried to come up with solutions, but it didn't make a difference. The company was not committed." Mary Lou Schaffer, president of the call center local in Pittsburgh, disagreed. "We were trying to work together. We had more say in the workplace as opposed to just grievances. It didn't take away from my time for union business." One AT&T executive estimated that WPOF saved the company hundreds of millions of dollars due to improved productivity, which was not a significant contribution, considering the size of the company. Another put the savings at $2 billion, including savings associated with not having to prepare for and operate strikes. Union leaders and managers appreciated the more open exchange of information that the WPOF structure facilitated and the opportunity it gave to increase their presence in the workplace, engaging with members in constructive projects. On the other hand, many managers resisted union participation in decision-making and were reluctant to transform unilateral quality initiatives into WPOF joint programs. Most business units did not participate in the program, and less than one-third of the AT&T workforce was involved at any level. Union participants on business planning councils found they often needed significantly more technical support, time, and resources to play a meaningful role. At the program's highest level, the human resources board did not meet regularly, and therefore the union did not realize its goal to have a greater voice in corporate strategic decisions.[26]

One of the most successful WPOF projects was the union-management initiative to transform the customer service operation in AT&T's BCS division. CWA used the process to win a substantial pay increase for the customer service employees in the business customer care organization, and as part of the job upgrade, succeeded in eliminating the second-tier wage rate for the account representatives in the billing offices, an unrealized union goal through several rounds of contract negotiations. In this section, I analyze why this particular WPOF labor-management program led to this positive outcome, one that proved elusive in the Bell Atlantic and US West call center job redesign projects that I discuss later in the chapter.

AT&T's 10 million business customers represented about one-half of the company's annual revenue, bringing in about $20 billion a year, with potential for enormous growth as new data services came online. In addition to competition from MCI, Sprint, and other long-distance companies, AT&T was preparing for the emergence of a new threat as federal legislation, which would open the door to RBOC entry into long-distance services, moved through Congress and eventually was enacted as the Telecommunications Act of 1996. In 1995, the AT&T BCS business unit launched a major reorganization initiative to create a more cost-efficient, streamlined customer service operation. The business unit leadership engaged the union in the project as part of Workplace of the Future. CWA selected Mary Jo Sherman, a union leader and frontline sales representative from the Syracuse, New York, call center, to work full time at AT&T headquarters as part of the team working on the customer service transformation. Ad hoc labor-management teams worked together on various aspects of the program. Union goals were incorporated into every component of the initiative, all training was jointly developed and implemented, and a union-management workforce planning subcommittee helped implement staffing changes and other workforce planning issues.

From the beginning, the customer service transformation project focused on creating value for three groups of stakeholders—customers, shareholders, and employees—by providing business customers a single point of contact for all sales, service, and billing functions. Before transformation, customers were required to call different service representatives depending on whether the issue was voice service, data service, or billing. AT&T was also looking to cut costs. The company's time studies showed that management sales teams were spending an average of 36 percent of their time performing post-sales work; if this work could be moved to union-represented workers, not only would the sales team have more time to spend on sales, but the company also could save money on labor costs. Also, by integrating customer service functions, AT&T believed it could reduce expensive redundancies and rework. According to William Stake, operations vice president for business customer care, "[I]f one person could handle a broader role, it could be done faster, better, cheaper." For AT&T's frontline employees, one-stop shopping would provide work with broader scope and variety, increasing their job satisfaction by empowering them to satisfy all their customers' needs.[27]

After several months of research, the customer service transformation team rolled out its proposal to reorganize work to provide one-stop shopping to customers, using different models depending on the size of the business customer and corresponding complexity of service needs. In each model, AT&T returned to a universal service representative job design that required cross-training to provide sales, service, and billing functions.[28] With the expansion of job responsibilities, AT&T created a new job title for the frontline call center worker: the customer service and sales specialist (CSSS). In response to union pressure, AT&T set the wage rate for this new position equivalent to that of the highest-paid occupational title in the business customer care organization. This represented a significant wage increase for employees who successfully transferred from the billing center titles to the new CSSS title, amounting to 9.7 percent (or about $3,200 a year) for billing inquiry representatives and 18.6 percent (or about $6,700 a year) for account representatives. In creating the CSSS title, CWA realized a decade-long goal to eliminate the second-tier titles in the business customer care organization. (The second-tier account representative title remained in the consumer call centers.) As AT&T Vice President William Stake remembers, "I got a lot of push back from my AT&T counterparts because in creating the new job we were actually . . . raising people's salary grade and pay. It was like 'you're crazy, you don't have to do this.' On the other hand, we were working with the union through this process." A labor-management workforce planning committee developed and implemented a training and bidding process for employees in the lower-paid billing center jobs to move into the upgraded CSSS position.[29]

The frontline call center workers and union activists were thrilled with the job upgrade. "The best part of customer service transformation was the wage increase," recalls Mary Lou Schaffer, who benefitted from the job upgrade. "It felt like the company valued you more. It took some time to get used to, but it was a much better time, feeling that we could really help the customer." The CSSS position was a complex job, requiring three and a half months of training. Mary Ann Alt worked in a center serving global customers and was assigned to the General Electric account. "They called me for everything, to make changes, and I was even writing technical orders." To assist employees, AT&T provided automated tools for access to customer records as well as information regarding AT&T products and services. Managers became coaches, helping the customer sales and service specialists improve on the job. While customer service transformation included some office consolidation, the number of union jobs grew. AT&T analysis found significant improvements in customer care operations from the reorganization.[30]

Several factors contributed to the project's success. First, there was significant overlap in management and union/worker goals to improve service, efficiency, and job quality by expanding the job of the frontline customer service employee as a single point of contact for customers. Second, the AT&T business unit management was willing to invest in a high-wage, high-skill worker because she served lucrative business customers; her ability to provide quality service and use her knowledge of the customers' needs to sell services returned significant value to the company. Moreover, transferring a portion of customer contact from more highly paid non-bargaining unit account representatives to the CSSS employee saved the company money. Third, the union was truly a partner in the process, with a full-time union leader working with the management team. The union ensured that workers benefited directly from the process improvements with a wage increase, strengthening employee support for the reorganization, which led to a more challenging job. Finally, the AT&T structure gave the business unit leader autonomy to run the business, which, in this instance, resulted in a successful WPOF business customer service transformation project.

Yet, the larger Workplace of the Future labor-management program ultimately proved unsustainable. The success of Workplace of the Future, like all union-management programs, depended on two key corporate commitments: first, a willingness to forgo short-term benefits to the bottom line, if necessary, to make longer-term investments in the workforce; and second, a recognition that corporate restructuring would not diminish union representation of the workforce. AT&T was not willing to make those commitments, slashing 36,000 union jobs between 1995 and 1999 and fighting aggressively to keep the union out of its growing wireless, cable, and consulting businesses. In 1995, AT&T initiated a major corporate and financial reorganization designed to boost shareholder value by selling off its NCR computer unit, divesting its manufacturing and installation business to form Lucent Technologies, and through the process, slimming down to its core communications business. Over the next four years, as AT&T purchased TCI, the nation's largest cable company, and grew its wireless business, union density plummeted to 27 percent in 1999. In 1996, CWA's Bahr warned that "the climate of trust that we struggled so hard to build . . . is rapidly eroding." In 1997, AT&T's new CEO, Michael Armstrong, announced another major cost-cutting initiative to reduce selling, general, and administrative (SG&A) costs from 29 to 22 percent of revenue by eliminating tens of thousands of jobs. By 1999, rather than perpetuate a charade of union-management participation, CWA formally withdrew from WPOF. Bahr expressed his disappointment: "Unions are ready to partner with management to improve productivity and employment security. First, however, we must overcome U.S. management's need to achieve short-term profits at any cost."[31]

With the demise of WPOF, joint union-management initiatives came to an end in the business communications service division. Mary Ellen Mazzeo, president of CWA Local 1152 in Syracuse and member of the BCS business unit planning council, reflected on the experience. "I don't believe we would have gotten the job upgrade without WPOF and the considerable buy-in by company executives," she recalled. But the experience also taught her important lessons about the limitations of labor-management

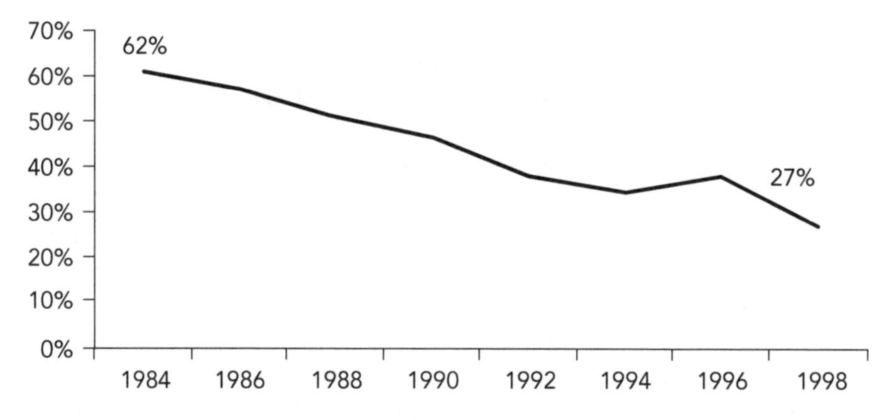

AT&T Union Density, 1984–1999.

programs in the context of the U.S. labor relations framework. "How naïve of us to believe that we really had a place at the table where union and company could work together in the best interest of the company, employee, and customer," she observed. "I think this exists in Europe. There may have been piece parts of WPOF where the union did well. But we deluded ourselves that we thought we had a real place at the table."[32]

Mega Teams at Bell Atlantic

In contrast to the Workplace of the Future contractually negotiated union-management program at AT&T, the Bell Atlantic Mega Team project to engage the union in redesigning the call center job was initiated by—and depended on—the commitment and leadership of one person, the vice president of consumer and small business services, Bruce Gordon. In this position, Gordon was the highest-ranking African American at the company. Bell Atlantic CEO Raymond Smith had just reorganized the $9 billion company into nine lines of business. He appointed Gordon, a twenty-five-year veteran marketing executive known for his maverick leadership style and willingness to work with the union, to lead its largest division, consumer and small business services. With its local monopoly in six mid-Atlantic states and Washington, DC, Bell Atlantic served virtually all 11 million residential households in its region, generating $4 billion in annual revenue, which represented 34 percent of total corporate revenue.

Anticipating more competition on the horizon, and influenced by the model of the UAW/GM joint program at Saturn, Gordon determined to improve the consumer division's performance by engaging the union. He reached out to the three CWA vice presidents responsible for the bargaining units at the Bell Atlantic local telephone companies and received their endorsement of the Mega Team union-management project. There was no formal written agreement between union and management regarding goals, structure, or resources. Although CWA previously objected to other Bell Atlantic unilateral employee participation programs that sidelined the union (known as the "Bell Atlantic Way"), in this instance, CWA leadership concluded that Gordon was serious about his commitment to full union participation. "Bruce Gordon was a friend of the service representatives," Sandy Kmetyk, president of the Pennsylvania call center local, recalled.[33]

Gordon invested significant resources in the Mega Team program, which was chaired by Gordon and a union counterpart, Barbara (Lephardt) Fox Shiller, assistant to the CWA District 2 vice president. All staff time, travel, and other expenses came out of Gordon's budget. While union and management cochaired all teams, management developed the meeting agendas and worked with others in the company to provide the data the teams needed to do their work. In other words, the union representatives depended on management for information and agenda setting, and they did not have independent resources to assist them in their work. For one year (1993–1994), fifteen union-management teams of seven to nine members each (for a total of more than 130 employees) met several times a month to address such core issues as job redesign, monitoring, staffing, training, measurement, and technology. In this section I focus on the job redesign team.[34]

The job redesign team, along with the other teams, began its work in August 1993 with an opening two-day conference in Philadelphia. In the union-only session, CWA district vice presidents Pete Catucci and Vince Maisano urged the union leaders to work with Gordon ("a union advocate"), while remaining "open and strong union people" by keeping a focus on union goals to expand union representation to all Bell Atlantic lines of business and remaining vigilant in contract enforcement. The consultant job redesign committee examined how best to integrate the billing, sales, and other functions of the new consultant job title. As discussed in chapter 2, the union had recently agreed to a new negotiated consultant job title that combined (and upgraded) the work of the lower-paid collections representative with that of the service representative's sales job.[35]

"When it first came about," Sue Anderson, union cochair of the job redesign team, remembered, "the union was surprised that the company would let us get involved to redesign the job. We were trying to make it more bearable, a place where you would want to go to work. I had high hopes and felt like the union and company could work together." Employee team members wanted to eliminate sales and adherence quotas, encourage teamwork, and have more time off the phones ("closed key time") to do follow-up paperwork, more training on new products and software systems, and more teamwork to allow greater job diversity.[36]

Similar to the AT&T job redesign team, the Bell Atlantic group aimed to design a one-stop customer service model, one that would eliminate costly errors and customer frustration from call transfers. But given the complexity of the job, which required knowledge of many different service packages, rate structures, and billing protocols, the team struggled to come up with a one-stop model. In the end, the team presented three alternatives. The first would have the consultant responding to all customer requests (sales and billing), the second would place the consultant in a co-located team with dedicated functions, and a third would cross-train the consultant in both sales and billing, rotating job functions, including time offline. The team never reached consensus on a recommendation. The job redesign team stopped meeting after September 1994 and the Mega Team project shut down a month later.[37]

Union minutes of team meetings do not record the reason for the shutdown, but clearly tensions had been building in the project's final months. In May 1994, union team members raised formal concerns that the partnership was breaking down, complaining that they were not getting the time they needed to devote to team projects. Managers were leaving the teams and opposed many of the recommendations. "We would make suggestions," Sue Anderson explained, "and the managers would say that it would require rewriting the computer program. They didn't want to do that. They wanted to get more sales without spending money on new computer programs." The company unilaterally implemented initiatives that undermined the Mega Team's work. There were reports of upcoming downsizing. Gordon responded to union concerns in a meeting with union team leaders. He remained committed to the project, he told them, but "it's harder than he imagined," as the teams tackled "tough issues of the business." By October 1994, Gordon's focus shifted with his appointment to the senior management team at Bell Atlantic. Without his support, the Mega Team program folded.

Anderson was disappointed but also emerged a stronger unionist from the process. "I met union leaders from throughout the Bell Atlantic region. I overcame my shyness and learned that I didn't have to be intimidated by management." Barbara Mulvey, president of the Salisbury, Maryland, local and cochair of one of the project teams, expressed her frustration. "Initially we thought we could achieve something. We proved that we could close our line to do paperwork or research to get answers for the customer and to give the consultant time to do her job. But management wouldn't work with us. They wouldn't buy into it, they were giving up control. Ultimately it was a joke." Local 13500 President Sandy Kmetyk took a different lesson, seeing the Mega Team experience as training ground for the future. "We came up with great ideas. Many went into bargaining proposals and the service representative relief package that we won after the strike in the year 2000."[38]

Why did the Mega Team project fail while the Workplace of the Future job redesign project in the AT&T business customer care unit succeeded? First, Mega Team, unlike WPOF, was not a union-negotiated program; it depended on the commitment of one individual, Bruce Gordon. When Gordon lost interest, the initiative ended. Second, the job redesign team was not able to reach agreement on a recommendation, although it is not clear whether the decision to shut down the project short-circuited final agreement. Third, the Bell Atlantic job redesign project was in the consumer line of business, whereas the AT&T project was on the business side of the company. Residential customers simply did not provide the same return to the company as business customers, and therefore Bell Atlantic managers did not want to make the significant investment in training and software systems that the job redesign required. Fourth, and most important, in August 1994 Bell Atlantic's corporate leaders announced its next phase of financial reengineering with aggressive cost reductions, including elimination of 5,600 jobs and call center consolidation, and adopted an aggressive stance toward the union. For example, when 1,200 technicians protested by wearing red t-shirts to work proclaiming "we won't be roadkill on the information highway," the company suspended them all.[39]

The union-management partnership in the consumer line of business fell victim to the companies' business strategy to boost earnings from its telephone business to finance investment in its wireless joint venture with NYNEX and network upgrades to enter the video business. This strategy positioned Bell Atlantic as a "telecom renegade" in 1995 contract negotiations when the company refused to follow the pattern set by the other six RBOCs in the triennial collective bargaining round. Relations between CWA and Bell Atlantic deteriorated as the union waged a mobilization campaign of workplace actions, outreach to shareholders, elected officials, and the public against Bell Atlantic, eventually reaching agreement on a contract five and a half months after expiration.[40]

The 1995 contract did not include any provisions to provide stress relief in the call centers. The positive labor-management relations fell apart, and the workers' hope to work with management to restructure the consultant job to provide the autonomy, support, training, and skills they wanted and needed to serve customers in a healthy environment was not realized.

Yet, the team members and the union had developed practical solutions that would resurface five years later as call center workers' demands for stress relief took center stage during the eighteen-day CWA strike against Verizon, which I discuss in chapter 6.

Job Redesign at US West

In 1993, the U.S. Office of Technology Assessment published a laudatory eighty-page report describing "a successful joint union-management effort to improve customer service and productivity at US WEST, Inc" in its consumer line of business. "[T]he company and the unions," the report noted, "have reached a series of mutual decisions that have protected union members' jobs and reorganized their work in a way that increases worker and customer satisfaction and that benefits the firm." That same year, President William J. Clinton chaired a panel at a Future of the American Workplace conference featuring the US West/CWA case study.[41]

Certainly, the union-management job redesign at US West seemed destined for success. US West was the smallest of the RBOCs, providing local telephone service in fourteen largely rural western states stretching from Iowa to Washington, with $10 billion in revenue in 1992. Union density remained stable at 62 percent, with 39,000 union-represented employees. Unlike the other RBOCs, the company and CWA maintained joint QWL teams in the years after divestiture. And, as an indication of company support, in 1989 they reached an agreement for company funding of fifty-eight "change agents" (half union-appointed, half management) to serve as internal consultants to support teams' joint work. A union-management Employee Involvement Quality Council (EIQC) composed of four union and four management representatives oversaw the joint projects. The job redesign, part of a larger union-management initiative in the US West consumer division known as Home and Personal Services, was governed by a memorandum of understanding that protected jobs and made a commitment to regard the union as an equal partner. The company paid for two skilled consultants—Kevin Boyle, chosen by the union, and Winnie Nelson, chosen by management—to facilitate the labor-management teams. Not only did the company pay for all team-related expenses, including time off the job and travel, but US West also covered the cost of union-only team meetings so the union representatives could discuss their own agenda before joint meetings with management.[42]

The origins of the job redesign project in the US West Home and Personal Services division began at the 1991 CWA national customer service conference. In a US West bargaining unit session, CWA activists shared stories about the highly stressful conditions in their call centers. The union leaders agreed to implement a district-wide mobilization plan to raise their concerns to US West management. When they returned home, they sent Jane Evans, the newly appointed vice president of the Home and Personal Services division, a bouquet of black balloons with a demand to change what they called "sweatshop" conditions in the call centers. To their surprise, Evans responded with an offer to form a labor-management team to "redefine the work" in the call centers that employed 5,600 people. She was well aware of serious problems in the division: there were large numbers of uncollected bills, the service representatives answered only half the calls within the regulatory sixty-second time frame, and there was high turnover and absenteeism. Flashing signs in the call centers frequently indicated 100 people were in queue waiting for someone to answer their call.[43]

Before the union would agree to the joint project, the four union members of the Employee Involvement Quality Council drafted a memorandum of understanding to guide the project, with provisions that guaranteed no layoffs, no downgrades, equal partnership, and respect for collective bargaining. To their surprise, Evans signed it. When the CWA locals unanimously ratified the draft, the project, dubbed "bunts and singles," took off. (Bunts and singles indicated that the team would work on many small initiatives.) The full team consisted of twenty-eight members, equal management and union. They initiated more than thirty projects to improve work processes. The team initiatives paid off in a $20 million reduction in uncollected bills, a 10-percent increase in customer access, and reduced absenteeism and turnover. To recognize collectors' importance to revenue generation, Evans upgraded the pay for the collector title to that of the service representative, a move that employees and the union applauded. (Like Bell Atlantic, US West had split the universal representative title into two job functions in the 1980s, downgrading the pay of the collectors who handled billing functions from that of the service representatives who handled sales.) The division added 250 positions in 1992. Most significant, Evans eliminated all performance appraisals early on in the process, convinced by management consultant Nelson that they did not improve performance.[44]

The team's most ambitious undertaking was the job redesign project. Linda (Armbruster) Mulligan, for the union, and April Hunter, US West labor relations manager, cochaired the team. Consultants Boyle and Nelson introduced the team to the socio-technical systems approach to job design, guiding them through a mapping process of the people (social system) and techniques, tools, and knowledge (the technical system) necessary to do a quality customer contact job. The team worked for a full year, frequently consulting with on-the-job customer service representatives. In December 1992, the team presented its recommendation to reorganize the consumer customer contact centers into cross-functional teams, with service consultants working in co-located, cross-trained teams, able to handle order taking and sales as well as billing and collections. The automatic call distributor (ACD) would be reprogrammed as a two-way system, not only directing customer calls to the appropriate call queue based on a voice response cue, but also giving the consultant information, in real time, so she could respond to call demand by switching the type of calls she would take. The service consultant's job would expand to include provisioning functions (establishing dial tone, assigning phone numbers), simple repairs, and the ability to talk directly to technicians in the field. Supervisors would become coaches who would use the information from the performance tracking systems to help service consultants improve. Peer coaching would be integrated into the team. Vice President Evans agreed to pilot the cross-functional teams in several call centers and selected Phoenix as the first test location. It was at this point that the Office of Technology Assessment published its report on the union-management project, of which job redesign was one part.[45]

Evans forwarded the job redesign proposal to her boss, Greg Wynn. He rejected the redesign as too expensive. In September 1993, US West CEO Richard McCormack announced plans to eliminate 9,000 jobs at the company, about 14 percent of the workforce, consolidate the call centers into mega centers in the larger cities, and close rural offices. These job cuts were on top of the downsizing of 6,000 positions announced two years earlier. Stock prices jumped 4 percent on the announcement, adding $470 million in value. US West opted to squeeze cash from the business to reward shareholders rather than support the longer-term investments in training and technology required to implement the recommendation of the joint union-management job redesign.

With the downsizing announcement, the authors of the Office of Technology Assessment report wrote, "This decision, which will reduce jobs in HPS [Home and Personal Services], underscores the fragility of high-wage, high-skill strategies. Although the unions will continue to work with HPS to improve work processes and redesign jobs, the partnership relationship has changed and the trust that had been developed has been undermined." In a report to a 1994 U.S. government conference, they put it even more bluntly: partnerships can't make progress when corporate decisions result in financial reengineering and indiscriminate job shedding.[46]

Job redesign union cochair Linda Armbruster Mulligan summed up her experience. "This was such a rewarding experience until you realized you spent all this energy for nothing. All that trust that developed after sharing your knowledge to come up with a proposal. In the end, we went back to an adversarial relationship at bargaining. We focused on getting things in the contract. There may be a place where you can combine the two, but it didn't work for us." The relationship between CWA and US West deteriorated, culminating in a 1998 strike over job standards, quality service, and health benefits.[47]

The union-management job redesign projects at AT&T, Bell Atlantic, and US West took place during a transitional moment for these companies as they transformed their managerial and financial systems from those of monopoly service organizations subject to regulatory oversight to those of profit-maximizing companies responding to the demands of financial capital. These former Bell companies faced real economic challenges competing with the non-union companies with lower labor costs and at the same time meeting capital demands to upgrade and invest in the latest technology. For a brief period, some managers at these companies reached out to CWA to work together to design high-wage, high-skill, high-quality work systems that would beat the competition. But the pressures of capital proved too strong as top-level corporate executives opted for cost-cutting reengineering over investments in high-performance work systems.

The experience at these telecommunications companies was not unique in the industry. Bell South, the regional Bell company in the nine southeastern states, experimented with self-managed work teams during this same period. The net impact was to increase workers' sense of discretion, satisfaction, and job security, but it had the opposite effect on first-line managers. Online teamwork shifted power from supervisors to workers, and the supervisors rebelled. Upper-level management abandoned a program that improved economic performance because frontline supervisors objected to the shift in power relations in the workplace.[48]

Contrary to the fears of Victor Reuther, Parker, Slaughter, and others, the Bell company customer service leaders and activists, for the most part, were not "co-opted" by management, but their hopes to turn participation programs into an expansion of collective bargaining to humanize work, stabilize employment, improve workers' living standards, and strengthen workers' collective power proved illusory, as companies opted for financial engineering to increase shareholder value rather than invest in their operations and their workforce.[49] The financial turn in management, with its focus on immediate cost minimization, short-run profits, and the price of company stock, sounded the death knell for these programs. With the corresponding decline in union power, management discovered that it could proceed unilaterally with restructuring initiatives that focused on financial reengineering and shedding employees.[50]

U.S. government policy provides few, if any, guardrails to prevent corporate leaders from reengineering the corporation to boost shareholder return at the expense of workers. In Germany and Scandinavia, codetermination statutes that mandate worker participation on corporate boards, worker/management works councils in the workplace, and sectoral bargaining that includes all firms in an industry sector strengthen union power and serve to block a race to the bottom in labor standards. Without such structural supports, U.S. firms that invest in their workforce for the long term often find they are punished by Wall Street, pressure that few corporate executives can resist and that unions alone cannot overcome.[51]

For the call center workers at AT&T and Bell Atlantic, financialization posed new threats to their employment security, compensation, and working conditions. With the declining cost and increasing capabilities of telecommunications networks, managers discovered that they could outsource call center functions to third-party vendors, thereby reducing the cost of labor and union-negotiated workplace rules, while demanding CWA concessions in employment standards and working conditions in the call centers to keep the work in house. The fight of call center workers and their union expanded to new terrain in campaigns against the outsourcing of their work to low-cost, non-union third-party vendors.

CHAPTER 5

Fighting for Job Security

"The company has hired and trained about 600 employees to steal our work," Linda Kramer, president of CWA Local 1023 in New Jersey, wrote to CWA regional Vice President Larry Mancino in June 1997. Kramer described her members' outrage over Bell Atlantic's outsourcing of over 200,000 high-value customer accounts to a Virginia-based non-union corporate subsidiary named Bell Atlantic Plus. "Suddenly our members are faced with declining call volumes. They are facing retribution for not meeting ever-increasing sales quotas while 'prime' sales opportunities are funneled-off to Bell Atlantic Plus. Bell Atlantic Plus is creating a very visible loss of work for our people. It's a real concern for many of them that we may be looking at layoffs."[1] Kramer, together with other CWA local presidents representing Bell Atlantic customer care representatives, urged Mancino to make the issue of call center outsourcing a top priority in upcoming contract negotiations.

By the end of the twentieth century, CWA customer service leaders and members faced a new challenge in their fight for good jobs as their employers began contracting out their work to third-party vendors. In their quest to cut costs and maximize flexibility, AT&T and Bell Atlantic managers adopted outsourcing strategies in their call center operations. AT&T outsourced more than half of its core residential long-distance customer service operation. Bell Atlantic contracted with the non-union subsidiary Bell Atlantic Plus to handle a new product line—internet sales—and to service high-value residential customers who purchased multiple services. Although CWA had long fought contracting out technician construction work, the outsourcing of customer service work—the face of the company to customers—was a new phenomenon. CWA and its call center members mobilized to block these outsourcing initiatives as a threat to employment security, collectively bargained wages, benefits, working conditions, and union power.

In this chapter, I analyze the economic origins of decisions by AT&T and Bell Atlantic to contract out call center work, the strategies that CWA and its call center members deployed to resist their employers' outsourcing, those strategies' effectiveness, and the origins of global offshoring of U.S. customer care jobs. Although both companies' outsourcing initiatives were a response to accelerated competitive forces and new opportunities resulting from passage of the market-opening Telecommunications Act of 1996, the driving force was the pressure from the owners of capital to boost shareholder value by slashing expenses and increasing market flexibility. CWA succeeded in beating back the Bell Atlantic outsourcing initiative, yet it barely stemmed the tidal wave of contracting at AT&T. Differences in product markets, financial condition, market competition, and union strength explain the different outcomes in the union campaigns to bring the contracted work back in house.

Financialization and the Fissured Workplace

The AT&T and Bell Atlantic contracting strategy is one piece of a monumental transformation and degradation of employment relations in the U.S. economy, one that labor economist David Weil calls the "fissuring" of the workplace. In the fissured workplace, a company or public agency sheds activities and services that were previously performed in house, and in so doing, reduces not only costs but also responsibilities connected to the employment relationship. The fissured workplace is a core component of the financial turn in U.S. capitalism and is now commonplace throughout the U.S. economy. In 2007, one-third of U.S. manufacturers' revenue and employment was produced at outsourced firms. The poster child for what economists call the "factoryless goods-producing firm" is Apple, which produces almost all its computers, tablets, and cell phones through subcontractors. To illustrate the pervasiveness of the fissured workplace, Weil describes the ways Marriott Hotels, Time Warner Cable, Bank of America, Walmart, and Hershey Company provide core functions, respectively, of hotel cleaning, cable installation, janitorial services, logistics operations, and candy production through a web of vendors, so-called independent contractors, and franchisees. "In an earlier era," Weil writes, "these and other large employers . . . would likely have directly employed the workers. Not so now."[2]

The fissuring of employment relations is a major source of the job insecurity, declining living standards, and gaps in the social safety net that have turned what were once good U.S. jobs into bad jobs. Fissuring drives a race to the bottom, as vendors compete with each other to win contracts based on lower labor and other costs. Responsibility for compliance with minimum wage and hour laws, and health and safety standards, or even ensuring that payroll, unemployment, and workers' compensation are paid are shifted from the lead business with economic resources and power over the relationship to contractors or franchisees who face pressures to skirt the law and depress compensation to keep expenses down. Contract and temporary workers, often misclassified as independent contractors, move from employer to employer, and frequently do not work long enough at one company to qualify for unemployment insurance or other protections.[3]

Certainly, capital owners' quest for cheap labor is not new. Labor historians have catalogued the story of capital flight by manufacturing firms throughout the twentieth century. Textile manufacturers moved south to avoid union labor; RCA opened and closed plants in Bloomington, Indiana, and sunbelt Memphis, Tennessee, before relocating to the maquiladoras in Ciudad Juarez, Mexico; auto companies moved first to the suburbs, then to the South and overseas; and General Electric CEO Jack Welch (nicknamed "Neutron Jack") famously announced that he would ideally locate his factories on a movable barge in a global search for low-cost labor and business-friendly regulations and exchange rates. Employer pursuit of non-union, low-cost labor drove location decisions by consumer electronics, meatpacking, steel, and countless other industrial companies, leading to decimation of rust belt communities and workers' living standards.[4]

At its core, then, the explosion of corporate subcontracting, franchising, third-party management, and outsourcing that began in the late 1980s and early 1990s was driven by the same goal to slash costs to increase value for investors. But the fissured workplace also represents something new. Rather than relocate their entire operations, many firms slimmed down to a very narrow set of what they considered their "core competencies," those that were the most profitable activities in the value chain. While advances in communications and information technology enabled this radical transformation of the employment relationship, it was changes in the nature of capital markets, combined with weak labor market institutions and unions, that allowed owners and shareholders to earn profits from financial engineering even as they frequently abandoned the actual production of goods and services through a web of outsourcing and franchising relationships.[5]

The fissured workplace represents the triumph of the financial over the managerial model of capitalism and the firm. In managerial capitalism, the firm's purpose is to produce goods and services to generate profits, and to reinvest retained earnings to improve productivity, profitability, and expand market share. Value is created, extracted, and distributed through the labor process; managers therefore rely on their workforce and share (to a greater or lesser degree) productivity growth with them.[6] In contrast, the firm's purpose in a financial model is to make as much money as possible for the company's owners, as measured by the share price. The financial model views the corporation as a collection of assets to be bought, sold, reengineered, and manipulated with the goal of increasing returns to shareholders. In the words of business historian Shoshana Zuboff, "the logic of capitalism [has] shifted from the profitable production of goods and services to increasingly exotic forms of financial speculation."[7] Capital is mobile and seeks the highest return, selling off those assets that are mature and less profitable, disaggregating and outsourcing to domestic or offshore contractors with lower wages and labor standards those activities in the value chain that generate lower returns. Top executives align their interests with shareholders through executive compensation tied to stock options. As a firm's financial success becomes less dependent on productive activity, managers increasingly view labor as another factor of production to be squeezed rather than a reciprocal relationship that could add value to the company. The logical extension of the financial turn is the outsourcing of larger and larger portions of the work producing goods and delivering services.[8]

U.S. labor market institutions have proved remarkably weak in countering the drive by capital to boost shareholder value through fissured employment structures. U.S. labor and employment law are based on stable employment relationships, and they provide few guardrails to prevent companies from offloading employment responsibilities, avoiding regulations, and downgrading labor standards through contracting arrangements. Unlike German and Scandinavian countries' sectoral bargaining structures, U.S. labor relations center on employer-based union representation, an institutional structure that facilitates the fracturing of what were once internal labor markets into core and peripheral employment. Unions face steep hurdles organizing subcontracted sectors. With notable exceptions, such as the SEIU's Justice for Janitors campaigns, few unions have developed successful organizing strategies that leverage power over the lead business to establish and maintain labor standards by their contractors. The National Labor Relations Board's shifting definition of "joint employer" complicates such organizing strategies.[9]

Given these organizing challenges, unions have focused on collective bargaining strategies to block or limit outsourcing, as well as legislative campaigns to raise the wages (and thus lower the incentives) to outsource work to low-wage, non-union workers.[10] In 1989, CWA negotiated agreements with AT&T and the RBOCs prohibiting contracting out of work that would lead to layoff, downgrade, or part-time work for bargaining unit employees.[11] These agreements have remained in the legacy Bell company contracts. But given the high turnover rates in the AT&T and Bell Atlantic call centers, the Bell employers' outsourcing strategies did not violate the contracts as the companies shifted jobs left empty through employee attrition to expand the work performed by outside vendors. The CWA call center worker activists mobilized to press their leaders to put new protections in place to ensure employment security for their customer service members.

The Telecommunications Act of 1996

The fight to block the AT&T and Bell Atlantic outsourcing initiatives took place in the context of two major developments in the communications sector during this period that profoundly affected the market structure, business strategies, union power, and therefore CWA's ability to bring the contracted work back in house. Congressional passage of the Telecommunications Act of 1996 heightened competitive pressures and unleashed new business opportunities for both companies, while the emergence of new internet and wireless telecommunications services offered areas for growth as well as substitution for landline voice telephony.

The 1996 Telecommunications Act, the first major overhaul of the 1934 Communications Act, in the words of its preamble, aimed "to promote competition and reduce regulation in order to secure lower prices and higher quality services for American consumers and encourage the rapid deployment of new telecommunications technologies."[12] With its emphasis on competition driving innovation and lower consumer prices, the 1996 Telecommunications Act was a quintessential reflection of Congress accepting neoliberal faith in the market to drive investment and consumer benefit.[13] The legislation opened all communications sectors to competition, ending the market segmentation of the original 1934 Communications Act and the lines-of-business restrictions of the 1984 AT&T divestiture decree. The legislation allowed AT&T to enter local service. It gave the Regional Bell Operating Companies (RBOCs), including Bell Atlantic, a path to enter long distance, contingent on opening their local bottleneck markets to competition. The legislation allowed telephone and cable companies into the others' former monopoly markets. Then Representative Edward Markey (D-MA), one of the legislation's main drafters, believed that competition, particularly between cable and telephone companies, would bring the benefits of the digital age to all Americans.[14] Ironically, the 1996 Telecommunications Act, designed to "reduce regulation," required extensive regulatory intervention to inject competition into the local telecommunications sector. The Telecommunications Act held out the carrot of RBOC entry into long distance but only after state and federal regulators certified, on a state-by-state basis, that the RBOC met a twenty-one-point market-opening checklist that included reconfiguring the network, developing automated ordering systems, and extensive rate-setting by state commissions.[15]

AT&T and Bell Atlantic (and other telecommunications companies) responded to the 1996 Telecommunications Act's regulatory restructuring with corporate reorganization, setting off a dizzying wave of mergers, acquisitions, and new business ventures. In 1997, Bell Atlantic merged with NYNEX, the regional Bell company in New York and New England; in 1999, Bell Atlantic began its entry into long-distance service; in 2000, the company combined with GTE, the largest non-Bell telecommunications company, to form Verizon Communications; and in 2006, Verizon fully integrated local and long-distance service with its purchase of MCI. AT&T divested its computer and equipment lines of business in 1996; bought and then sold a cable company in 1999 and 2002, respectively; launched a local service business in 1998; spun off its wireless subsidiary in 2002; closed its local service operation due to regulatory changes in 2004; and agreed to acquisition by SBC Communications, a former regional Bell company, in 2005.[16]

During this same period, the development of the commercial internet and auction of additional wireless spectrum spurred AT&T, Bell Atlantic, and other telecommunications companies to invest in and market internet and wireless services. Bell Atlantic used cash flow from its dominant local voice service to pay for digital networks and wireless technologies, while AT&T regarded its declining consumer long-distance service as a source of cash to bolster earnings as it tried to reinvent itself as an all-distance voice, video, and wireless company.[17]

In this hyper-competitive yet uncertain environment, Bell Atlantic call center managers looked for strategies to handle the sale and servicing of new products and more complex services, while AT&T customer service leaders looked for ways to cut costs in a rapidly declining consumer long-distance business. Management at both companies strove to meet demands from investors to boost shareholder value by keeping a firm rein on expenses, particularly labor costs. For the AT&T and Bell Atlantic customer care managers, outsourcing by contracting with third-party vendors offered attractive solutions to these business challenges. Call center vendors paid lower wages and few, if any, benefits, which lowered the largest variable cost (labor) of their customer service operations. Moreover, the contractual relationship could be structured to expand and contract as demand changed, providing flexibility to respond to fluctuating and unpredictable call volumes. Finally, contractors' employees do not show up in the employment numbers reported on financial statements, sending a positive signal to investors. The downside of the vendor relationship, however, was loss of control over the service quality, particularly when provided by a low-wage, higher-turnover workforce; the added expense of contract management; the risk of poorly written contracts; and dependence on the outside vendor for knowledge of the costs and best practices of running the business.

Unlike most U.S. workers, call center employees at AT&T and Bell Atlantic had a union to organize a fight against the fissuring of their jobs through outsourcing strategies. The long struggle of the majority-female CWA customer service leaders and members to elevate their issues on the union agenda finally paid off as top union leaders made their battle against outsourcing a high priority. In the AT&T bargaining unit, the 10,000 CWA-represented call center workers, concentrated in about a dozen large locals, comprised about one-fourth of all union employees at the company, where union density had fallen to 27 percent in 1999. Union density at Bell Atlantic remained high throughout this period, with strong customer service locals in Pennsylvania and New Jersey. The 1998 Bell Atlantic merger with NYNEX, with its militant union-represented New York workforce, doubled the number of union-represented workers at the merged Bell Atlantic to almost 100,000, with union density at 69 percent.[18]

Losing the Battle against Outsourcing at AT&T

By the end of the twentieth century, the AT&T consumer business was in a struggle for survival, as market share and revenues in its core consumer business—the saturated long-distance voice market—tumbled and the company struggled to enter the local telephony and consumer internet businesses. In 1999, the AT&T consumer line of business brought in $21.7 billion; five years later in 2004, consumer revenue plummeted to $7.9 billion, a 275-percent decline. The sharp drop was due to a combination of factors, including substitution from wireless and e-mail, declining long-distance rates, and most important, RBOC entry into long-distance service, beginning with New York in late 1999 and expanding to every state by the end of 2003. The RBOCs found it relatively inexpensive to offer long-distance service and quickly expanded into this market. For example, just

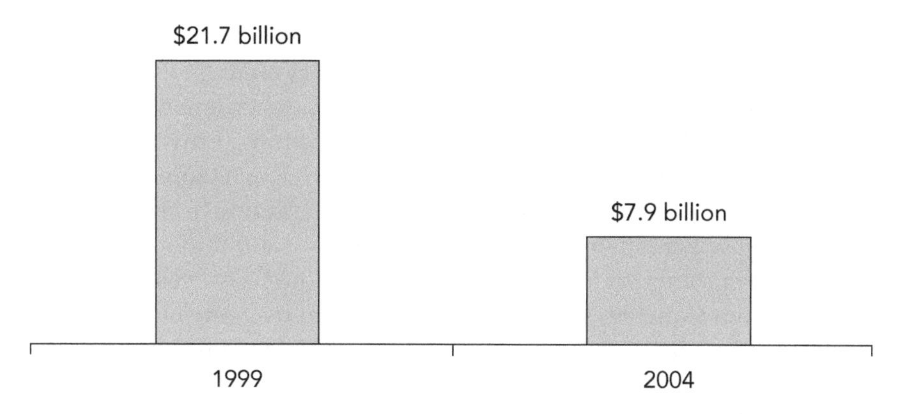

AT&T Revenue from Residential Customers.

one year after Bell Atlantic entered long-distance service in New York, a full 20 percent of customers had signed up for its long-distance service, and by 2004, almost half were long-distance subscribers.[19]

In contrast, the economics did *not* work for AT&T entry into local service. It was too expensive to build local networks to reach every customer location across the country. Therefore, AT&T (and the other long-distance companies MCI and Sprint) provided local service through a resale strategy, leasing network access from the local RBOCs. Even though state and federal regulators initially established favorable regulated leasing rates, AT&T (and the other long-distance companies) were never able to capture a significant portion of the local service market. In 2004, when AT&T chose to abandon its venture into local service, the company had captured only 4.2 million subscribers to its local/long-distance service bundle and only 1.2 million DSL internet customers. The Telecommunications Act proved a boon to the RBOCs once they were allowed into long-distance service (in the 1999–2002 period), but a bust for the long-distance companies that could not break into the local market.[20]

Facing these financial challenges in its consumer line of business, AT&T adopted a fissuring strategy by outsourcing a portion of its long-distance customer service operation beginning in 1999. The frontline customer service representatives in the AT&T consumer call centers, with the job title of account representative, began to see more and more notations in customer records of calls handled by outside contractors, and local union leaders saw employment decline in the consumer call centers, with a corresponding drop in the number of union members in their locals. The account representatives noticed that a company called Convergys was a lead vendor. Convergys's roots go back to the period right after divestiture, when AT&T set up a call center in Jacksonville, Florida, to answer shareholders' stock ownership inquiries. Within a few years, those calls declined, and AT&T turned the call center into a telemarketing subsidiary named TransTech (also called AT&T Solutions) that contracted in work from other firms, with primary staffing through temporary agencies. In 1998, AT&T sold TransTech to the telemarketing firm Convergys. The purchase agreement required AT&T to buy $300 million in annual customer care services from Convergys over the next eight years (through 2006), giving Convergys the right of first refusal over any other vendor. Therefore, as AT&T consumer managers looked to save costs, outsourcing to Convergys filled a significant portion of the company's $300 million subcontracting obligation.[21]

Because AT&T's consumer customer service operation operated on a national call distribution system, any union effort to reduce the company's outsourcing had to be addressed by the national union. Therefore, as CWA members in the call centers pressured local union leaders to do something to protect their jobs, they in turn elevated the issue to Ralph Maly, CWA vice president with responsibility for the nationwide AT&T bargaining unit. Maly had limited options for fighting the outsourcing, since the CWA/AT&T contract banned contracting out work only if the practice caused actual layoffs or turned full-time into part-time work. Because of high turnover in the call centers, AT&T reduced the number of employees through attrition and did not hire new employees to backfill the vacated positions; the contract language therefore was not applicable.[22]

Maly reached out to William Stake, vice president of AT&T sales and customer care, with demands to stop the contracting out of CWA members' call center work. Stake was a thirty-year veteran at AT&T with years of experience working with the union, most recently as the manager in charge of the successful labor-management Workplace of the Future business customer care transformation project discussed in chapter 4. Stake proposed, and the union accepted, creating a union-management consumer sales and service center (CSSC) subcontracting committee to explore mechanisms to "maintain and grow bargaining unit work while maintaining the competitiveness of AT&T Consumer." Stake believed that the consumer business unit may have gone too far in subcontracting out so much of the consumer call center operation. As Stake later explained, "we were headed in the wrong direction, there was too much emphasis on cost reduction and we were losing sight of the customer. I was trying to balance these two things: we want to preserve the experience for the high-value customer, and the low-value customers are going to have to suffer in order to bring down the unit cost in total."[23]

Stake and CWA Vice President Maly chaired the CSSC subcontracting committee. Committee members included union leaders from the consumer call centers and managers who worked in Stake's consumer care operation. Three of the six CWA members of the committee were women, local presidents representing the largest AT&T consumer call centers in Dallas, Texas, Kansas City and Lee's Summit, Missouri, and Mesa, Arizona. AT&T provided the committee with detailed information and analysis regarding call volumes, cost, quality, and employment at the in-house and outsourced call centers, as well as funding for the union to hire its own financial expert, Randy Barber. The committee functioned over a four-year period, from its initial meeting in 2001 until 2005, when SBC Communications purchased AT&T (and renamed the merged company AT&T).[24]

At the first committee meeting in August 2001, AT&T presented a detailed report on CSSC subcontracting. At that time, CWA represented 5,259 call center employees working in eleven consumer centers across the country, down from 7,554 employees two years earlier, a loss of more than 2,150 in-house positions.[25] CSSC subcontracting had grown substantially over this two-year period, from 830 full-time equivalent (FTE) employees to 4,204 employees in 2001. If AT&T had kept the work in house, union employment in the CSSCs would have grown, not diminished, by about two thousand positions. A full 65 percent of long-distance calls were handled by fifteen U.S.-based vendor call centers. The AT&T in-house centers took all the more complex calls regarding the emerging local service business, but since the long-distance call volume dwarfed local service, the outsourced call centers handled a full 50 percent of all calls to the AT&T consumer business.[26]

AT&T explained to the committee that the main metric the business used for cost comparison was a cost per call calculation, determined by taking the total expense allocated to a call center divided by the number of calls handled by that center. According to this metric, the cost per call at the internal (in-house) sites exceeded those of the external (vendor) sites by as much as 34 percent. AT&T predicted that the gap would widen to about 40 percent by the end of the year as the company renegotiated its vendor contracts. AT&T highlighted high absentee rates, averaging about 14 percent among in-house call center employees, with a trajectory to cost the in-house channel $16.1 million in 2001. In contrast, vendors did not have absentee expenses, since their employees were not entitled to sick pay and could be fired at will, with no union contract to protect them.[27]

Union committee members challenged AT&T's cost per call calculation with two major objections. First, they argued that the comparisons were not apples to apples because the in-house call centers handled more offline paperwork than the external centers, thereby increasing the cost per call in the in-house centers. Second, the union believed that in-house employees provided better quality service, which could be quantified and could narrow the cost differential with the external centers. AT&T conducted further analysis into these issues. The company calculated the annualized costs to the business of the lower vendor service performance at $44 million.[28]

But even with the $44 million quality adjustment, outsourcing delivered a significant savings to the business during this period of declining call volumes and revenue. While the union representatives felt some vindication that the analysis provided solid documentation of the in-house quality difference, this was not enough to convince AT&T to bring the low-value long-distance calls back in house. CWA asked its consultant Barber to conduct further analysis, adjusting the unit cost differential for the higher rates of offline paperwork performed at the internal centers. AT&T gave Barber detailed site-specific data to use in his analysis. Barber's analysis found that employees at internal centers spent 17 percent more of their work hours doing offline paperwork than employees in external centers did. Adjusting the unit cost differential for both higher internal site quality and more time spent offline, Barber calculated a 17 percent to 20 percent cost per call differential between internal and external centers. This certainly closed the cost gap somewhat, but not enough.[29]

To close the remaining unit cost gap, CWA presented a detailed proposal to AT&T. The union was on the defensive, essentially negotiating from a position of weakness. The proposal indicates how far the union was willing to go to protect jobs, offering concessions that would have been unacceptable in a more stable period. CWA offered to accept a lower-wage, temporary job classification to handle customers with low monthly charges, but only if the company agreed to set a baseline of permanent employees. CWA also suggested adopting a team-based incentive pay plan to reward sales and good attendance. In addition, CWA recommended creating a union-company committee to redesign the customer service job, which the union suggested would help reduce absenteeism.[30]

AT&T's Stake rejected the union's insourcing proposal because it did not, in his view, provide the savings that he needed to remain cost competitive. "We are still very far apart on closing the expense gap that exists between the internal and external sites," he wrote to CWA's Maly. He rejected the suggestion that increased internal call center volumes or the second-tier wage proposal could reduce the cost differential, which he expected would widen in 2002 due to renegotiated vendor contracts. Because the union proposals did not close the unit cost gap, Stake explained that he could not "justify bringing work into the internal centers." However, the AT&T vice president responded positively to the other three union suggestions and agreed to create subcommittees to address absence and wellness policies, job redesign, and incentive pay proposals. "But please understand," Stake warned, "that until we can quantify the benefits these initiatives might bring to the business, we cannot include them in our unit cost gap analysis."[31]

Over the next three years, union-management subcommittees worked together to develop and analyze the results of various initiatives in job redesign, wellness policies, and incentive pay. The subcommittees designed pilot projects and evaluated the impact on sales, quality, absence, and unit cost. While the pilots showed some improvements, and the front-line account representatives especially liked the job redesign, the results were not sufficient to overcome AT&T management's laser focus on cost as call volumes continued to drop rapidly. The results of the trials did not convince AT&T to insource more work.[32]

Closing the unit cost gap became even more challenging after December 2002, when Convergys convinced AT&T to route calls to Convergys call centers in New Delhi and Bangalore, India, where labor costs were significantly lower than at the domestic outsourced centers. The cost savings were significant, and within three years, 42 percent of AT&T calls were sent offshore to call centers in India, Canada, Mexico, the Philippines, and Panama.[33]

The transition from domestic to offshore outsourcing marks an important moment in the evolution of the call center industry, in general, and the customer service operations of U.S. telecommunications companies. The process was driven by the call center vendor Convergys in its search for cheap labor to gain competitive advantage over other call center companies located within the United States. Other vendors adopted the practice, giving birth to what is today a global call center industry. The nature of the industry both facilitates going offshore and poses unique challenges. On the one hand, in contrast to factory production, opening a call center requires few sunk capital costs, giving call center vendors great flexibility in location decisions. On the other hand, customer contact work requires language skills and cultural understanding—challenges that can frustrate customers and that continue to plague offshore call centers. In those early days of global offshore outsourcing, AT&T was willing to subject its low-revenue customers to these potential frustrations because the cost advantages were so appealing.[34]

AT&T call volumes continued to plummet. In just three and a half years, from summer 2001 to January 2005, AT&T consumer call volumes declined by 300 percent, from 5.4 million to 1.5 million calls per month. By 2005,

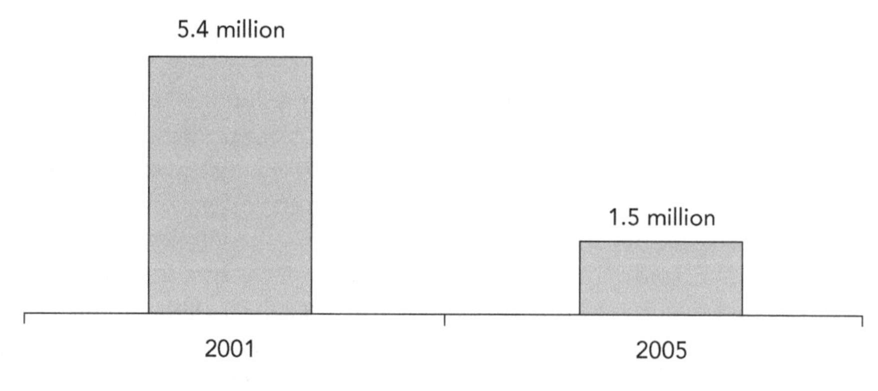

Monthly Calls to AT&T Consumer Call Centers, 2001 and 2005.

AT&T had closed several internal call centers, including the Mesa, Arizona, mega center that once employed several thousand account representatives, and reduced the internal CSSC workforce by more than two-thirds.[35]

In the end, CWA could not convince AT&T that the company could fulfill the CSSC's mission to "grow bargaining unit jobs" during a period in which the company failed "to maintain the competitiveness of AT&T Consumer." AT&T retained a roughly 50/50 ratio in its distribution of calls to internal and external call centers throughout this period. The union could claim a partial, though still bittersweet victory; as call volumes declined, AT&T reduced the number of calls sent to vendors, to maintain the 50/50 internal/external ratio.[36]

In the years after Congress passed the 1996 Telecommunications Act, AT&T management failed to adopt a winning corporate strategy to take advantage of new growth opportunities. AT&T management squandered resources on a failed venture into the cable business, buying and then selling its cable division to Comcast three years later.[37] AT&T sold off its wireless business, the fastest-growing sector of the communications industry. In the end, AT&T saw no path to success, and on January 1, 2005, SBC Communications, one of the former regional Bell companies with a CWA-represented workforce, agreed to acquire AT&T for $16 billion. November 18, 2005, signaled the end of an era for what was once considered "the biggest company on Earth." CWA local union leaders took to wearing a T-shirt with the slogan "AT&T: We built it, management destroyed it."[38]

Blocking Outsourcing at Bell Atlantic

In contrast to the experience at AT&T, CWA largely succeeded in blocking Bell Atlantic's initial foray into outsourcing its call center work. Beginning in the fall of 1997, Bell Atlantic consultants (the service representative title Bell Atlantic adopted a few years earlier) notified CWA local leaders that they observed notations in customer records that third-party vendors were performing some of the work of CWA-represented employees. The contractors were selling long-distance service and performing disconnect functions at wages as low as $6 per hour, with few benefits. The union leaders were especially concerned about an entity called Bell Atlantic Plus, a non-union Bell Atlantic subsidiary with a large call center in Hampton, Virginia. The company opened the facility in the summer of 1997 with more than 700 employees, with plans to grow by another 200 workers. According to a Bell Atlantic press release, the center was designed to provide one-stop shopping to customers for local voice, wireless, internet, and (upon state approval) long-distance service. Bell Atlantic explained that "Bell Atlantic Plus is a direct outcome of the Telecommunications Act of 1996 which opened the local market to competition. Freedom to choose brings with it enormous complexities for the busy consumers. . . . By integrating our services, we intend to make telecommunications shopping simple and easy."[39]

CWA union leaders and activists were incensed by the creation of Bell Atlantic Plus, particularly in light of the company's rejection a few years earlier of the job redesign proposal discussed in chapter 4 that would have paved the way for the one-stop shopping that Bell Atlantic now planned to locate in a non-union subsidiary. In the fall of 1997, several hundred predominantly female CWA activists from around the country who were in Williamsburg, Virginia, for the annual CWA customer service conference rallied outside the nearby Bell Atlantic Plus call center. Their message to company executives: the fight to shut the subsidiary down and bring that work in house was a top union priority.[40]

As CWA prepared for 1998 contract negotiations with Bell Atlantic, the first since the 1996 Telecommunications Act and merger with NYNEX, the union made gaining union representation over Bell Atlantic's growing wireless, internet, and long-distance lines of business a top priority. CWA leaders understood that this was a critical moment to fight for union jurisdiction over the new lines of business. While 80 percent of Bell Atlantic revenue came from local service in 1998, the union was well aware that this would shift in the future as wireless and digital technologies eclipsed landline voice telephony. This was the moment for the 100,000 union members at the new merged Bell Atlantic to join together to mobilize in a contract fight to "bargain for the future," the name CWA leaders gave to its 1998 contract campaign. The CWA education and mobilization manual explained that "Bell Atlantic is pursuing a strategy of union-containment—isolating the union to representing workers in the traditional 'Plain Old Telephone Service' (POTS) segment of the business, while new fast-growing services such as wireless and Internet will remain non-union." The manual explained that if Bell Atlantic's strategy of union containment succeeds, "CWA will represent an ever-shrinking percentage of Bell Atlantic's workforce, and the union's power to negotiate good contracts will weaken, making it harder and harder to protect our standard of living."[41]

The CWA education and mobilization manual included projections for wireless and internet growth, and it also highlighted the threat that Bell Atlantic Plus posed to the future of the union's customer service members. This represented a significant development, reflecting the growing importance and political power of the largely female customer service members within the union. CWA had long opposed contracting out technician work; now it added opposition to outsourcing call center jobs. The manual told the story of a Bell Atlantic accounting center closing in suburban Philadelphia. Even though several Bell Atlantic subsidiary work locations were nearby, the displaced union members were not allowed to transfer to them. "Had those subsidiaries been organized," the manual recounted, "workers would not have been faced with painful choices, like the single parent with 28 years' seniority who had to uproot her family and follow her job or else lose her eligibility for full pension benefits."[42]

In the summer of 1998, CWA leaders received even more troubling news about Bell Atlantic Plus from Melissa Morin, president of Local 1400 in New England and a longtime national leader of the CWA customer service annual conference. Morin reported that Bell Atlantic was now transferring the accounts of its high-value customers to Bell Atlantic Plus. CWA-represented consultants in the Bell Atlantic call centers would no longer handle these accounts, even for traditional voice telephony, long the sole jurisdiction of union-represented consultants. This development meant that Bell Atlantic was moving its most lucrative customers, those who purchased multiple bundled services from the company, to Bell Atlantic Plus. This posed a major threat to the CWA-represented workforce in the call centers.[43]

In 1998 contract negotiations, CWA President Morton Bahr made the fight to block the outsourcing of work to Bell Atlantic Plus a top priority. In a July 1998 letter to Donald J. Sacco, Bell Atlantic executive vice president for human resources, Bahr wrote that the Bell Atlantic Plus issue "must be resolved if a contract of any duration is to be reached peacefully. There is just no way that we will ever consent to moving our work to another subsidiary with lower wages and benefits." Moreover, he argued, "it is just bad business strategy. When the company makes mistakes we lose jobs.

. . . [T]here is an industry consensus that a bundled bill is very valuable. . . . But placing your best customers at the mercy of high turnover, low paid workers doesn't make much sense. It would make more sense to serve these premier customers with your best, most experienced employees." Bahr cited evidence from a CWA survey that showed that not a single other RBOC had adopted this strategy. "US West already tried this," he wrote. "They now have the work performed in the core company and estimate they have a 35 percent overall savings."[44]

As the contract expiration deadline approached on August 8, the parties were still far apart. CWA members from New England to Virginia went on strike. The walkout lasted only two days, leading to a significant CWA victory not only on wages and benefits, but also, in the words of the CWA press release announcing the end of the short strike, "extending union representation over jobs on the information highway." In a bargaining report on the settlement, CWA leaders noted that the contract's "crown jewel" was the provision that ensured CWA members would perform all work related to internet, DSL, video services, alarm monitoring, and long distance. Bell Atlantic agreed to transfer all Bell Atlantic Plus accounts to CWA-represented call centers by March 30, 1999. The company also agreed that CWA-represented employees would sell and service all bundled items that include services, such as voice telephony, that had historically been serviced or sold by CWA-represented employees.[45]

The 1998 CWA and IBEW mobilization of its tens of thousands of members in support of the contract campaign and the two-day strike came at a time when Bell Atlantic was eager to avoid an extended labor confrontation. Having recently announced its $65 billion merger with GTE Communications, the largest non-Bell telecommunications company, Bell Atlantic was focused on regulatory approval in twenty-seven states, at the FCC and the Department of Justice. In this context, the company was willing to concede to the union's demands for jurisdiction over all services delivered over its wireline network.

There was one gaping hole in the 1998 "jobs of the future" agreement. The company refused to extend neutrality/card check recognition to Bell Atlantic Mobile, the company's fastest-growing subsidiary. (Under a neutrality/card check agreement, the company agrees to remain neutral in any organizing campaign and to recognize the union when a designated percentage of employees sign cards certified by a neutral third party in support of union representation.) According to CWA President Bahr, Bell Atlantic Vice Chairman Ivan Seidenberg appealed personally to Bahr to hold off until the 2000 contract negotiations on the demand for wireless organizing rights. At that time, Seidenberg, heir apparent to become

CEO of the new merged Bell Atlantic, committed to Bahr that he would negotiate a card check/neutrality agreement for wireless. According to Bahr, Seidenberg asked for time to "get his arms around the company." Bahr was sympathetic. He had a long relationship with Seidenberg, dating back decades when Seidenberg began his career as a New York Telephone technician and Bahr was a local union leader. "We reached a verbal agreement," Bahr later explained. "I was agreeing to take Bell Atlantic out of the [three year] bargaining cycle of the other companies. [And I agreed] that we don't make any demands for card check [in 1998]. The quid pro quo [was that] we would get it in the year 2000."[46] Winning organizing rights at Bell Atlantic (renamed Verizon) wireless would become a major strike issue in the next round of bargaining two years later.

The market-opening Telecommunications Act of 1996 intensified competitive pressures in the telecommunications industry, driving call center managers at both AT&T and Bell Atlantic to adopt fissuring policies to cut costs, gain flexibility, and avoid adding to their employment head count. The 1996 Telecommunications Act proved a death sentence for AT&T and other long-distance companies, as the barriers to entry into local markets proved too steep to replace the precipitous drop in long-distance service. CWA simply could not prevent AT&T's race to the lowest-cost solution as the company struggled, ultimately unsuccessfully, to survive. The union's twenty-year battle to save good jobs at AT&T ended in November 2005 when SBC bought what was once the iconic AT&T.

In contrast, CWA succeeded in blocking Bell Atlantic's foray into call center outsourcing and winning jurisdiction over all but wireless lines of business before competition eroded the union's substantial membership strength in what was still the company's core business, local telephony. Yet, the fight for job security at what would become Verizon Communications after the GTE merger would continue as the merged company's focus on maximizing shareholder value continued to press call center workers to do more with less. CWA customer service members and the local leaders who represented them were prepared to make stress relief a top priority in the 2000 round of bargaining, and they discovered a sympathetic public would support them even as they walked off the job to win good working conditions in the call centers.

CHAPTER 6

Striking for Stress Relief

On August 6, 2000, 87,000 workers from Maine to Virginia walked off the job in an eighteen-day strike by CWA and IBEW against Verizon Communications, the new company formed from the recent merger of Bell Atlantic and GTE. "I thought our three-week strike against Verizon in August 2000 was the greatest strike ever," Marilyn Irwin later recalled. "I loved how people in different job titles cared about each other's issues. Technicians spoke with consultants on the picket lines and declared that they didn't want to return to work until the consultants' issues were addressed." In her position as local union vice president, Irwin represented hundreds of customer service representatives at Verizon's large call center complex in suburban Maryland. She marveled at the unity on the picket line and the support the predominantly male technicians gave to the bargaining demands of the majority-female service representatives.[1]

CWA and IBEW members walked the picket lines in the summer of 2000, not over economic disputes, but over three key issues: stress relief in the call centers; organizing rights at Verizon Wireless, the fastest-growing segment of the company; and employment security. *The New York Times* dubbed this a strike in which an "old economy" telephone workforce took a stand for "new economy" issues—high tech wireless jobs, stress in the automated call centers, and employment security in an age of global capital mobility. When the strike ended, the unions claimed victory on all three fronts, with a card check/neutrality agreement covering organizing at Verizon Wireless; strict contractual limits on transferring work from the East Coast to lower-cost locations in the national Verizon footprint; and most relevant to this study, a stress relief package that the union and its customer service members hailed as a high point in their struggle to win humane conditions and job security in the call centers.[2]

The successful strike of 87,000 Verizon workers in nine states and Washington, DC, in the year 2000 stands as a significant achievement.[3] Building and maintaining solidarity among two different unions (CWA and IBEW) that were negotiating contracts for seven different regional bargaining units organized into seventy-one diverse union locals posed particular challenges.[4] Scholars have documented a precipitous decline in union strike activity since the early 1980s. Labor work stoppages are down from an annual average of 303 large strikes involving 1.5 million workers (25.5 million lost days of work) in the 1947–1979 period to an average of 50 strikes per year involving 325,000 workers (7.3 million work days) in the 1980–2006 period.[5] Even the well-publicized 2023 strike wave, which involved 27 work stoppages by 447,000 autoworkers, actors and screenwriters, nurses, teachers, hotel employees, and others who walked the picket lines for a total of 15.6 million lost work days that year, pales in comparison to the number of strikes that took place in the years before President Ronald Reagan quashed the Professional Air Traffic Controllers (PATCO) walkout in 1981. Although the 2023 strikes were noteworthy for the significant victories the workers won, most U.S. strikes in the previous decades largely failed, even those conducted with militant tactics and broader social vision.[6]

How, then, were the CWA and IBEW workers and their union able to defy this historical pattern to win a work stoppage against the tenth-largest corporation in the United States, one with $65 billion in annual revenue?[7] And how were the call center workers and their union able to turn resistance to the stressful conditions at work into lasting change? Both the structure of the telecommunications industry in which Verizon operated in the year 2000 and the high level of internal union education, mobilization, and organization were critical to the work stoppage's success. The stars aligned: the strike took place at a time when union power was strong and Verizon was vulnerable to the negative impact of strike activity. As for the service representatives, their many years of networking, organizing, and running for local union offices paid off as they had amassed substantial power and influence within their union. Their issues—single mothers being forced to work last-minute mandatory overtime with no one to pick up their children from day care and workers who could not take the time needed to respond to customers' needs—tapped into the general public's anger over poor customer service and frustrations in finding work and family balance. Finally, with the introduction of competition into the local communications market in the wake of the Telecommunications Act of 1996, the customer service representatives occupied a strategic position critical to the business's success. All this came together in the 2000 strike.

Yet, while the stress relief agreement went further than any previous settlement in protecting workers from the worst managerial abuses in the call centers, it left in place the basic work organization and technology systems that allowed the company to control the labor process and the pace of work. Thus, this pathbreaking stress relief package also reveals the boundaries of collective worker power in the contested terrain of the highly automated call centers operating within the weak labor market and regulatory institutions of the U.S. political economy.

Mobilizing and Organizing to Win the Strike

CWA leaders recognized that union power was at a crucial turning point at Verizon in the year 2000. (The CWA and IBEW bargaining units involved in this strike were from the Bell Atlantic portion of the merged and renamed Verizon.) The company still dominated the provision of local telephone service in its geographic footprint, earning more than 67 percent ($43.3 billion) of corporate revenue and 60 percent ($10 billion) of operating profits from the wireline segment of the business. Having doubled in size four years earlier with the Bell Atlantic/NYNEX merger, the union represented virtually the entire occupational workforce on the wireline side of Bell Atlantic, with 69 percent union density.[8] But CWA represented only about 100 of the 8,000 wireless workers, the fastest-growing segment of the company. Bell Atlantic aggressively resisted all organizing of wireless workers. CWA leaders understood that unless Verizon wireless workers won organizing rights, union power at the company would decline as wireless eclipsed the wireline segment of the business. Moreover, winning organizing rights at Verizon Wireless was unfinished business from negotiations two years earlier. As discussed in chapter 5, CWA and Bell Atlantic agreed to a card check/neutrality agreement in 1998 that excluded Verizon Wireless. At that time, CWA President Morton Bahr believed he had a verbal commitment from CEO designate Ivan Seidenberg to extend card check/neutrality organizing rights to Verizon Wireless workers in the 2000 round of bargaining. In 2000, CWA expected Seidenberg to deliver on that promise.[9]

Union leaders calculated that the time was ripe to take a stand in 2000. The telecommunications sector was a vibrant domestic industry not subject to global competition. Verizon had not yet begun sending call center jobs offshore. Most important, in June 2000, two months before the unions walked off the job, Bell Atlantic and GTE consummated their merger, taking the name Verizon Communications to become the largest telecommunications company in the United States providing local, long-distance, internet, and wireless service. The unions struck just as Verizon initiated a branding campaign to win and retain customers. Verizon did not want the bad publicity associated with a strike. The unions took advantage of this strategic moment—timing was all important.

But timing and market conditions were not everything. The unions faced many challenges winning a successful strike and sustaining unity between two unions involving multiple bargaining units with workers in hundreds of work locations employed in diverse occupations with varied priorities. Success required a keen strategic assessment of union power, ability to win public and political support, identification of deeply felt worker issues, a militant union culture, extensive preparation, member education, an effective mobilization and communications structure, skilled negotiators, strong and creative leadership, a keen sense of timing, and a measure of good luck and fortune. CWA and IBEW pulled all this together in 2000.

CWA leaders and activists had sustained a long tradition of strike action based in a deeply rooted and widely shared view that failure to reach agreement at contract expiration would lead to work stoppage. In 1983, 700,000 CWA members struck for twenty-two days at the Bell system, and in 1986, CWA members walked off the job for three weeks at AT&T. The legacy of militant strike action was particularly strong in the Northeast. In 1971, 40,000 CWA members at New York Telephone struck for 218 days over the demand for national bargaining. Participation in what the union dubbed the "Spirit of '71" strike became a "badge of honor," according to strike veteran Christopher M. Shelton (who served as CWA president, 2015–2023). In 1989, 60,000 CWA and IBEW members in New England, New York, and New Jersey walked off the job at NYNEX Communications for seventeen weeks under the rallying cry "Health Care for All, Not Health Cuts at NYNEX." To this day, CWA members wear red shirts every Thursday in memory of Edward "Gerry" Horgan, who was killed in that strike when a manager's car plowed into a picket line. Red shirt Thursdays are just one way that veterans of the 1989 strike kept the militant spirit alive, passed down to the next generation of union activists and leaders. Even in the historically less militant Bell Atlantic-South bargaining unit, which included right-to-work Virginia (along with Maryland, West Virginia, Pennsylvania, Delaware, New Jersey, and Washington, DC), workers and the union had developed a more combative posture with a work-to-rule campaign in 1995 and a two-day strike in 1998.[10]

The unions built solidarity through a comprehensive program of education, communication, and mobilization that was implemented through a structure of local workplace leaders. Early on, CWA and IBEW agreed to joint bargaining, blocking the company's ability to play one union against the other. One year before contract expiration, the unions began preparing a coordinated education campaign and a systematic program to rebuild a mobilization and communication structure in every workplace. The education program was delivered by 1,500 trained union activists and focused on the national union's top priority to win organizing rights at Verizon Wireless. The widely distributed *Bargaining for Our Future* mobilization manual explained that "[t]he union will lose power and leverage if management is successful in its attempt to seal us in the slow growth basic telephone sector and keep us out of high growth/high revenue services such as wireless. We must organize wireless at Bell Atlantic . . . to maintain our power and the ability to protect our jobs, wages, benefits and working conditions. The union must win CARD CHECK AND NEUTRALITY in order to have a decent chance of organizing wireless workers at Bell Atlantic because current labor laws are stacked against workers." Also, the union bargaining agenda included issues that deeply affected every member, including demands to reduce excessive mandatory overtime, protect job security, and obtain stress relief in the call centers. Members engaged in an escalating series of activities before contract expiration to build unity and test the mobilization structure.[11]

When the contract expired on August 5 with no agreement, the months of preparation and building of internal structures paid off. Almost every worker walked and stayed off the job throughout the strike. CWA and IBEW maintained unity with picket duty (newspaper accounts cite as many as 14,500 strikers on 300 picket lines at any one time), more than fifty "children's rallies" ("I'm walking a picket line with my mama"), and an additional twenty-five rallies on missed paydays, ranging in size from fifty in small locals to more than 12,000 at Verizon headquarters in New York City. The unions reached out to political and community leaders; to cite one of many examples, Senator Chuck Schumer spoke at the New York City rally. Activists formed "flying squadrons" to follow managers who were sent out on repair and installation calls. "Mobile picketing is going strong. Only 25 percent of Verizon's trucks are even out on the streets," one upstate New York local reported. A Brooklyn local recounted following 80 percent of the seventy-five trucks the company sent out, while totally shutting down six of the sixteen garages in the borough. A Verizon spokesman told *Newsday* that mobile picketing was "preventing [managers] from doing their work." In liberal New York City, strikers garnered significant public support, with customers calling the union to report scab locations. Union leaders and activists stayed on message with the press, and nearly daily media coverage quoted workers explaining that "wireless needs to be unionized so we can keep the jobs" and "my family doesn't want to get uprooted any more than yours does."[12]

Picketing service representatives told reporters about the pressures they faced on the job. "There's no downtime," Patti Egan, a customer representative in New York City, said. "The customer disconnects and the next call's right there. Try living like that, taking calls every two seconds." She explained that she had to finish the paperwork from one call even as she took an order on a new call. Sometimes, she said, she couldn't finish the paperwork. This added to the stress since serving the customer was "the inner part of the rep." Service representatives connected the stress of the job with the need to win organizing rights at Verizon Wireless. "Forget about the stress of my job with things like random monitoring of my calls by a supervisor or forced overtime," Stephanie Harris, a service representative in a New York City suburb, said. "I'm worried that I won't have a job in five years because wireless will be dominant and plain old telephone service will be very small."[13]

Local union leaders and staff reported daily on strike activity and sent pictures of picket duty and rallies to the national union, which were redistributed to union members in a daily strike bulletin and on a strike website. From Richmond, Virginia, came the report: "The Children's Rally today was a great success! Approximately 300 to 350 red-dressed members along with their children chanted, sang songs, and picketed for about 1 hour." The Brooklyn, New York, locals reported a "huge rally with 1,650 adults, 110 kids, two clowns, and a 30-foot rat. A crowd of about forty kids stood with clenched fists, shouting at managers looking out the window, and leading the chant 'We want our mommies' and daddies' jobs back.'" The West Virginia locals sent in a copy of an ad in the local paper that a state representative placed in support of the strike. Another West Virginia local proudly exclaimed that "our flying squadron went out today" and sent managers away from a big job that brings in $25,000 per month in revenue. From Virginia Beach, Virginia, came the report that "yesterday's RALLY FOR RESPECT . . . was electrifying. Over 300 members formed an ever-growing picket line." A rally in Boston's City Hall Plaza created "a sea of red . . . and black . . . of striking IBEW and CWA and community supporters from all over New England in a crowd of 3,000." Local unions of teamsters, teachers, steelworkers, paperworkers, painters, carpenters, longshoremen, and state and local labor federations joined rallies and picket lines. Locals competed for bragging rights on their rallies, picket lines, and mobile squadrons, reinforcing a sense of pride, accountability, and solidarity. The events were covered widely in the press.[14]

Strike activity engages every member who walks off the job and creates opportunities for creativity and building leadership. Barbara Wago, a union steward from the customer service Local 13500 in Pennsylvania, sent President Bahr an exuberant email describing her first time speaking in public at a local payday rally in Harrisburg, Pennsylvania. "I have never before today given a public address," she wrote. "I had so much adrenaline flowing . . . because I believe what I was saying. I told the press, news media & approximately 200 union members that we are here today in 100% support of our negotiating team. That today is our last paycheck until we have respect . . . job security, the right to organize . . . a fair contract. Our stress levels reach all-time highs."[15]

The union walkout could not shut down the fully automated Verizon communications network. Customers continued to make phone calls and transfer data throughout the strike. The replacement workers, many of whom were managers given emergency training to climb poles or access customer account databases in the call centers, struggled to do the frontline workers' jobs. While the strike exacted an economic toll on Verizon's business—there was a reported backlog of at least 280,000 repair and installation orders by the end of the strike—the workers and their unions won the strike through their unity in moving public and political opinion, which convinced the company to concede to the unions' top demands.[16]

Negotiations over the wireless card check/neutrality language consumed the attention of top union leaders and Verizon negotiators. Fifteen minutes before contract expiration at midnight on August 5, Ivan Seidenberg, then Bell Atlantic CEO and future Verizon chief executive, called CWA President Morton Bahr with the message that the company would accept card check/neutrality at Verizon Wireless. The two leaders delegated negotiation over the specific language to Larry Cohen, then CWA executive vice president (and future union president), and Jack Navarro, Verizon's labor relations director, and Marc Reed, vice president for human resources at Verizon Wireless. Lawyers from Jones Day, a notorious antiunion law firm, assisted Verizon in the negotiations. Over the next twelve days, as union members walked the picket lines, Cohen and the Navarro/Reed team exchanged multiple drafts of a proposed memorandum of agreement regarding card check/neutrality at Verizon Wireless.

Early on, CWA and Verizon reached agreement on a card check process and set a recognition threshold at 55 percent of a bargaining unit. Thus, the major point of contention was over the definition of corporate "neutrality." Cohen initially proposed neutrality language lifted directly from the 1998 Bell Atlantic agreement that covered all Bell Atlantic subsidiaries except wireless. That language was identical to the neutrality provisions in the 1997 CWA/SBC agreement, under which tens of thousands of SBC wireless workers had joined CWA free from employer intimidation, as discussed in chapter 3. CWA's proposed neutrality language required the company to remain silent in any organizing campaign and to hold its managers accountable to refrain from any communication regarding the union. The language barred the company and its managers from expressing "any opinion for or against Union representation" and from any statements regarding "the potential effects or results of Union representation." It also required the company to instruct managers on this definition of neutrality.

Verizon insisted on taking out this language. Verizon, on the advice of their Jones Day lawyers, countered with a definition of "neutrality" that would give the company and managers the right "to express themselves on the 'pros' and 'cons' of Union representation" and "to respond openly to employees' questions" about the impact of union representation. While recognizing that this language could open the door to employer interference, Cohen could not get the Verizon team to budge and eventually accepted the weaker neutrality language. On August 18, the thirteenth day of the strike, the parties inked the memorandum of agreement regarding card check and neutrality with Verizon Wireless, with the provision that it would expire in four years. At the time, Bahr hailed the agreement as "an important breakthrough . . . that secures the future of our members at this company." In retrospect, CWA's Cohen realized that the company's insistence on management's right to express their views regarding union representation indicated that Verizon never really intended to abide by corporate neutrality at Verizon Wireless.[17]

The regional Verizon-North (New York and New England) and Verizon-Mid-Atlantic (New Jersey south to Virginia) bargaining tables reached final agreement on all other issues on August 20 and August 23, respectively, claiming victory on priority concerns regarding employment security, jurisdiction over jobs of the future, and stress relief in the call centers. The union won representation over internet work and limitations on mandatory overtime. Verizon agreed that it would not move more than 0.7 percent of jobs annually in any occupational group beyond state boundaries, a major guarantee of employment security for all workers and especially for those in the call centers whose work could easily be rerouted to other locations. This protection effectively prevented AT&T-style consolidation of call centers and contracting out of sales and service jobs, preserving the small- to medium-sized call centers and local jobs throughout the Verizon east coast footprint. The three-year agreement included a 12-percent wage increase, many other economic improvements, and virtually no givebacks. Most relevant to this study, the settlement included the commercial stress relief package that was the top priority for the call center segment of the union.[18]

Building Power for a Stress Relief Package

By the year 2000, working conditions in the Bell Atlantic call centers had reached the breaking point. Serious understaffing at a time of growing customer demand drove managers to intensify speed-up, raise sales quotas, increase supervisory monitoring of worker performance, impose many hours of mandatory overtime that interfered with child care and family time, and block workers' ability to transfer into non-sales jobs. Workers were angry and demanded change. The challenge for the union activists and leaders was to turn this anger into organized collective action that would result in contractually enforceable improvements.[19]

After many years of networking, organizing, and running for local union offices, the predominantly female customer service segment of the union had amassed substantial power and influence within CWA. (CWA represented almost all of the customer service employees at Verizon. The IBEW represented technicians in New Jersey and technicians, operators, and a small unit of service representatives serving business customers in New England.) There were about 17,000 call center workers across the Verizon East Coast footprint, composing about 25 percent of the union-represented labor force. Union activists in the call centers were now powerful leaders of large customer service locals in Pennsylvania, New Jersey, New England, and metro New York, and in many of the amalgamated locals in the other states. Some were appointed to staff positions and were top assistants to their regional vice presidents. These call center leaders had worked hard over the years to educate their regional vice presidents (all former technicians) about the customer service issues and to develop relationships with the largely male technician workforce that still dominated the union. They had developed strong connections with each other at CWA conventions, district meetings, and the annual customer service conference. In addition, these customer service leaders had a fierce advocate at the very top of the union in Dina Beaumont, a former operator whom we met in earlier chapters. Beaumont, now executive assistant to President Morton Bahr, took responsibility in 2000 to coordinate negotiations over customer service issues, making sure that the unions would not resolve the strike until the call center leaders were satisfied.

The leaders of the local unions representing customer service workers, with assistance from the CWA research department, began preparing many months before bargaining to identify problems and develop a common agenda. CWA research department staff interviewed customer service leaders and in January 2000, the CWA research director distributed a memo on "Service Rep Issues for Bell Atlantic Bargaining" to the three district vice presidents responsible for Verizon bargaining, the bargaining team chairs, regional mobilization coordinators, and Beaumont. While the bargaining list would be refined over the next months, the memo details the major problems the call center workers and their leaders identified.[20]

The first problem was serious understaffing. After years of downsizing, there simply were not enough employees in the call centers to meet customer demand for second lines for dial-up internet and local/long-distance service bundles. Understaffing in the call centers created a vicious cycle in which managers intensified speed-up, mandatory overtime, sales quotas, and monitoring. This in turn led to even more demanding conditions in the call centers, accelerating turnover and low morale among those who remained. Even when authorized to hire more consultants, the company could not train and retain enough new employees for the complex customer service work to replace those that were leaving due to retirement, transferring to less stressful positions, or simply quitting. In one large New Jersey customer service local, for example, the number of consultants was down 25 percent (from 1,300 to 970) compared to two years earlier. In Pennsylvania, the collections department's head count was down from 900 to 300. According to data provided by the company to CWA, job tenure had dropped precipitously in the call centers. In the two largest customer service locals in New Jersey and Pennsylvania, almost one-quarter of consultants had been on the job less than one year. This was not due to a recent increase in new hires but rather to high turnover rates. By the year 2000, the job had become so challenging and stressful that only 19 percent of new hires in the Verizon Mid-Atlantic region and only 31 percent in Verizon North were still on the payroll after one year. (These high turnover statistics stand in stark contrast to those eight years earlier. In 1992, fewer than 8 percent of customer service workers in the Pennsylvania and New Jersey customer service locals had less than one year job tenure.) As Melissa Morin, president of CWA Local 1400 in New England and one of the CWA customer service network leaders, wrote in a CWA message distributed at a Verizon shareholder meeting in March 2000, "A truly good service rep takes four to five years to develop. But they're treated so badly that they're bailing out, and the customers are suffering. Right now, it's happening in the residential consumer offices. When it builds on the business side, Bell Atlantic is going to start losing their high-value customers. They'll be losing millions and millions of dollars."[21]

Second, with fewer consultants trying to serve growing demand, managers imposed mandatory overtime, often at short notice, to get the work done. The company imposed required overtime of as much as two hours, particularly on busy Mondays. Consultants with children had to scramble to cover childcare, particularly when they had little advance notice. Others had to miss school and family time on mandatory Saturday workdays. "Verizon makes us work at least 10–15 hours of overtime each and every week," Cleo Young, a consultant from New York, announced in a CWA radio spot. "I want to spend more time with my four-month-old son and my husband, but Verizon forces us to choose between our jobs and our families."[22]

Third, there was no "closed key" time off the phones to do follow-up work. In prior years, consultants typically received a small amount of time on their schedule to stop receiving incoming calls so they could follow up on unresolved customer issues. They cherished this "closed key" time, as they called it, not only because it provided a welcome break from the constant customer contacts but also because they felt an obligation to meet their commitments to customers. Not being able to meet those commitments added to the stress at work. Due to staffing shortages, Verizon eliminated any closed key time, even in Pennsylvania, where the union had negotiated fifteen minutes closed time on Tuesday through Friday, subject to the needs of the business. Verizon now determined that business needs required eliminating closed key time in Pennsylvania and in all the call centers.

Fourth, to keep consultants in the call centers, Verizon did not allow consultants who were not meeting their sales objectives to bid to transfer to open positions in non-sales jobs. The company also increased the period of time that a consultant had to work in the call center from twenty-four to thirty-six months before submitting a transfer request. Because the vast majority of consultants were not making their sales objectives (17 percent in New England, 37 percent in Virginia), these policies effectively closed down transfers out of the call centers. The perverse consequence was that consultants who were not good salespeople were frozen in sales jobs (unless they quit).[23]

The call center leaders continued to refine their agenda at a CWA bargaining council meeting in March 2000. The union typically convened a bargaining council several months before contract negotiations were scheduled to begin. Hundreds of local leaders, including significant representation from the customer service units, were in attendance to set bargaining priorities and to prepare the member education and mobilization campaign. The national union distributed a background briefing paper prepared by the CWA research department entitled "Justice on the Job: Working Conditions in the Customer Service Occupations." The intended audience included the union members and leaders from all occupational groups at the meeting, their members back home, and Verizon negotiators whom the union knew from past experience would be keeping a close eye on union preparations for bargaining. The briefing paper articulated the union's vision for the call centers. "Understaffing and stressful working conditions make it difficult to provide good service, which adds to the stress. The key to success in this competitive era is quality service, which derives from the skill, expertise, and morale of the workforce. Management by stress undermines employees' ability to provide high quality service, with serious implications for employee well-being and Bell Atlantic's bottom line. . . . Bell Atlantic must invest in customer service and recognize the full value that each residential and small business customer represents." The "Justice on the Job" paper identified four ambitious priority areas for bargaining on customer service issues that should be familiar by now to the reader: first, a demand to increase staffing levels; second, an end to secret monitoring, adherence, average work times, unrealistic sales quotas, and a guarantee of closed key time every day; third, relaxation of force freezes and time-in-title requirements to transfer; and fourth, protections against mandatory overtime. At this meeting, the delegates adopted a formal bargaining resolution that included the call center workers' priorities, further cementing the entire union's commitment to relief in the call centers.[24]

Recognizing the complexity involved in negotiating solutions to address working conditions and understaffing concerns in the call centers, Verizon and the unions agreed to begin discussion on these issues in early June before the opening of formal negotiations. CWA leaders from all the large customer service locals across the Bell Atlantic footprint, as well as staff who were former service representatives, convened in June to prepare the union agenda for the discussions. Seven of the nine CWA local leaders were women, as were two of the four CWA staff, reflecting the emergence of strong female leadership among this occupational group.[25] They developed a sixteen-point proposal outlining what "our members need to relieve stress," indicating their unified support for concrete solutions for their members. In addition to items that I have already identified, the list added several items related to job security, including the return of subcontracted work and guarantees that all future sales and service work would be given to bargaining unit employees.[26]

The company/union joint meetings began in early June and continued for ten days, reconvening for two days in mid-July after formal bargaining began. After mutual agreement that the churn rate in the call centers was unacceptable, Verizon and the unions reached agreement on only one issue—limitations on monitoring. This agreement required advance notification of monitoring and placed limits on the number of monitoring sessions based on the consultant's job performance. In reaching early settlement on monitoring language, the company and CWA drew on earlier pilot projects trialed five years earlier during the Mega Team project discussed in chapter 4 and on language negotiated previously by the powerful customer service locals in their New Jersey contract.[27]

But the company and union could not reach agreement on other issues. Verizon proposed adding a commission sales plan, a regional call-sharing queue that would distribute incoming calls across a wider geography, and extending the standard work week from thirty-seven and a half to forty hours. The union rejected all these proposals. "Everything you have brought is retrogression," CWA District 13 staff representative and former service representative Jim Byrne told the company representatives. On the last day of the meetings, union chair Pat Niven told her counterparts that the company's proposals did not address the stress problems. The meetings adjourned on July 16, three weeks before contract expiration. Union participants informed their regional vice presidents and Beaumont about the lack of progress.[28]

The bargaining teams continued negotiations over the stress relief package. On August 1, 2000, the Mid-Atlantic bargaining unit made a formal presentation to the company negotiators with the opening line that "[t]he abuse and stress that Verizon has placed on our consultants/service reps . . . has been abominable, and our Union can no longer allow these terrible working conditions to continue." The union presented the company negotiators with a list of sixteen issues to be addressed. The list included mandatory overtime, closed time and a period of delay between incoming calls, transfer rights, sales quotas, training, internet work, subcontracting, flex time, monitoring, adherence, appraisals, and wages. When the contract expired on August 5, the union and the company were still far apart on these issues.[29]

The next day, on August 6, when CWA and IBEW called the strike, the call center workers joined the work stoppage and the union structure went into overdrive coordinating strike activity. "Once we were on strike, we had picket captains in addition to the mobilization coordinators," Victoria Kintzer, secretary-treasurer of the statewide Pennsylvania local, recalled. "We were on the picket lines in front of the business offices. The picket captains would set up the picket duty schedule, make sure people were there, keep up morale on the picket line. If we had any scabs [among the Pennsylvania unit of 3,000 consultants], it would have been a handful. We stayed in close touch with the technician local, so if we had problems [with scabs], they would help us." Other union leaders recall few consultants who crossed the picket lines, even in right-to-work Virginia.[30]

The customer service workers mobilized to make sure that the union would not settle the strike without agreement on their concerns. "Dear Mr. Bahr," Deana Smith wrote in an email on the strike's fifth day, "I'm sure you know that our strike vote [in our office] was 95% . . . [t]he stress of the job is beyond belief. I know more people on Prozac than not in the pittsburgh [*sic*] rssc [residential sales and service center]. Please do something to alleviate the stress." Kim Rogers, consultant and union steward in a residential center in Ardmore, Pennsylvania, echoed these sentiments in her email to President Bahr. "I am sending this E-mail to make sure that both myself and the other consultants are not forgotten during bargaining. Our office had a strike authorization vote of 100%. We work long hours and are subject to constant monitoring by humans and computers, and very rarely are we able to receive the scheduled close time. . . . We need some relief and are hoping the bargaining committee will be the ones to help us get it." President Bahr responded to these emails: "Please tell your colleagues that all of us here in bargaining are standing strong to resolve the commercial marketing issues . . . we are as strong as we were on the first day of the strike."[31]

CWA press releases, radio spots, and flyers designed to win public support for the work stoppage emphasized the stressful conditions in the call centers and the work/family dilemmas resulting from mandatory overtime. A CWA flyer explained that "we're on strike" to demand "the elimination of needless job stress and harsh working conditions" in the call centers. CWA released a press statement headlined "Too Much Stress Means Workers Suffer—And So Does Customer Service." CWA ran radio ads in ten markets featuring two consultants talking about forced overtime and stressful working conditions. "My job is helping customers and solving problems, but the pressure at work makes it awfully hard," consultant Marilyn Irwin from Maryland told listeners. "This strike is all about dignity and respect. We're taking a stand for ourselves, our families, and our customers."[32]

The press picked up on the call center issues. *The New York Times* ran a major story headlined "When 'May I Help You' Is a Labor Issue" that detailed the difficult working conditions in the call centers. In the article, the reporter noted that "novel though it is for stress to be an issue in a strike . . . many Americans can understand complaints about work spilling over into family time" and the toll that emotional labor takes on one's health. The reporter explained to her readers how the automated monitoring systems followed the service representative's every move. "Nearby the supervisor's stations, a computer screen shows every workstation on a color-coded grid. The squares change color depending on whether the employees are signed on, signed off, keeping a customer on hold, and so on. A supervisor can see at a glance if a representative is taking too long and go investigate." Dawn Barbour, consultant from Madison, New Jersey, explained that she must follow a Verizon script in conversations with customers, regardless of the context, which makes her feel like "an idiot." Mike Karas told a Pittsburgh reporter that monthly sales quotas had increased from $8,000 five years ago to $60,000, with a company requirement to make a sales attempt on every call, no matter the circumstances. "There are times when you don't have a second between calls. It's one right after another," he said. The *Pittsburgh Post-Gazette* featured an article about life in the call centers headlined "Are Service Reps Facing Last Call for Joy on the Job?" and *The Washington Post* asked "Home Life on Hold? Verizon Strikers Say Call Centers Take Toll."[33]

At CWA headquarters, Beaumont kept in close touch with the call center leaders and bargaining teams, maintaining unity, assisting in identifying the key demands, and crafting bargaining solutions. "I want to thank you for working so closely with President Bahr, our bargaining team, and local officers of Districts 1, 2, and 13," Linda Kramer, president of Local 1023 in New Jersey wrote to Beaumont during the strike. "The insight and coordination of all involved, I know, will bring us success in this round of negotiations." Kramer went on to praise the important benchmarks established at the call center leaders' meetings before bargaining, which "brought all the Commercial/Marketing representatives together in a spirit of true unity that we did not have in the past."[34]

On August 20, the Verizon North bargaining unit reached agreement on the unions' key demands, and three days later, on August 23, the Verizon Mid-Atlantic unit resolved several local issues and inked a tentative agreement. (The agreement was tentative pending approval by all members of the bargaining unit. Members overwhelmingly approved both settlements.) Examining the final August 23 stress relief package as compared to the company's offers before and during the strike reveals significant concessions by Verizon. The company dropped its demands for a sales commission compensation plan and call sharing across state boundaries, a move that the union believed would have opened the door to call center consolidation and job cuts at smaller call centers. Verizon agreed to provisions that reduced the intensity and pace of work for the consultants and effectively required the company to adopt policies to hire, train, and retain more consultants. These included a guarantee of thirty minutes daily of closed key time (except on busy Mondays), limits on mandatory overtime, and more liberal transfer rights from sales positions.

The members also won important employment security protections, bringing back contracted disconnect work, and a commitment to train all consultants to sell and service high-speed internet, making union-represented in-house call centers the primary channel for this work. Verizon had resisted this latter provision, caving only when it became clear that it was necessary to settle the strike. The settlement agreement also included the monitoring language from the pre-bargaining discussions, requiring advance notice and limits on the number of annual monitoring sessions (twenty for highly rated consultants, thirty for those meeting all requirements, and forty for those needing improvement). Finally, the consultants won a 4-percent wage adjustment on top of the general 12.5-percent increase.

Union leaders from the customer service units reveled in their achievements. "We did one hell of a job and we were so proud of what we won," Victoria Kintzer, union leader from the Pennsylvania local, recalls. "The members appreciated the union," Carol Summerlyn, staff representative and former consultant in Virginia, remembers. "They understood that the union bargained some relief for them."[35]

The 2000 strike for stress relief represented the pinnacle of years of organizing, networking, strategizing, and building power by the call center workers, activists, and union leaders. This all came together in the year 2000. The onerous working conditions in the call centers created a vicious circle of high stress, high turnover, speed-up, and intensification of monitoring—which exacerbated the stress and induced more turnover. Verizon managers recognized that they had to make improvements, and the union and members pushed forward with an agenda for change. They built unity among themselves and with their union brothers and had a fierce advocate at the top of the union structure. They drew on lessons learned over the years, contract provisions negotiated in one contract, and experiments piloted during the earlier union-management Mega Teams projects (discussed in chapter 4). They contributed to, and benefited from, the militancy of their union technician brothers. They won support from the public. The year-long union-wide education and mobilization preparation and the workplace structures paid off when 87,000 workers from both unions walked off the job for organizing rights, jobs of the future, and stress relief in the call centers. The timing was right: union power was still strong in the dominant wireline business, and the newly merged company did not want a strike to tarnish its image with the public.

The union leaders and members were justly proud of the provisions that they won in the stress relief package that curbed the most abusive speed-up, surveillance, mandatory overtime, and contracting of the members' work. These limitations made a concrete difference in the call center workers' daily lives. They were especially proud of the guaranteed thirty minutes per day off the phones, which relieved stress in at least four ways. It gave them a break from nonstop customer interactions; it slowed down the pace of work; it brought variety to the job; and most important, it allowed them to follow up on commitments to their customers, providing the quality service that was central to their identity as customer service employees.

But the stress relief package was silent on fundamental issues regarding work organization and control over the automated systems that structured the pace and manner in which they did their work. Absent broader public policies that provide workers, through their union, a real voice in technology deployment and work organization, the call center workers and their union leaders at Verizon were not able to wrest more fundamental control over the conditions at work. The workers' unity and determination could not overcome the structural and institutional barriers that limited their voice over the very nature of work in the digital workplace.

Epilogue

In the face of tremendous odds, the majority-female union-represented call center workers won important victories. They built power within their union. They negotiated meaningful constraints on the worst uses and abuses of automated technology, giving them some measure of control over the manner and pace at which they did their jobs and protections against intrusive surveillance of their every movement at work. They won a degree of recognition for the value their skills, knowledge, and emotional labor created, with periodic wage increases that exceeded the general wage improvements negotiated for all workers in their bargaining units. They won union jurisdiction over jobs of the future in internet sales and service. They mobilized to win neutrality and card check recognition that enabled AT&T wireless workers to join CWA free from fear and intimidation. They went on strike to win a stress relief package at Verizon. They built solidarity despite the walls that divided them in their isolated cubicles.

And yet, union power proved necessary but not sufficient to address the root causes of the insecure, unhealthy, and dehumanizing jobs in the call centers. The contest between the CWA-represented call center workers and their employers took place at a time that neoliberal ideology, with its overriding faith in unfettered market forces, provided few, if any, public policies to restrain the power of capital. Telecommunications policies favored non-union new entrants. Weak labor laws were no match for aggressive anti-union employers determined to prevent their workers from organizing. Enterprise-based bargaining provided few avenues for union voice in deployment of technology and investment decisions. Financial deregulation eliminated the few public policy guardrails on capital, both reflecting and contributing to the financial turn in capitalism and management practices that drove cost-cutting, outsourcing, intensified speed-up, and stressful surveillance in the call centers. The call center workers and CWA wrestled with employers for control in an arena that was heavily stacked against them.

This study ends in the year 2000 with the heady strike victory at Verizon, in which the workers and their union won neutrality and card check recognition at Verizon Wireless and a stress relief package in the call centers. The card check/neutrality agreement proved to be worth less than the paper it was written on, cementing the worst fears of the CWA negotiators who accepted the weak neutrality language to end the strike. Verizon refused to comply with the agreement and violated it with impunity. During the agreement's four-year lifespan, the company shut down three of the four Verizon Wireless call centers in the Northeast, shifting the jobs to right-to-work (for less) states in the South that the agreement did not cover. The company set up an anti-union website, held mandatory anti-union meetings, and retaliated against and fired union activists. When the card check/neutrality language expired in 2004, Verizon refused to renew it. Without employer neutrality, few Verizon Wireless workers were willing to risk threats and intimidation to seek union representation under the National Labor Relations Act's inadequate protection.[1]

Verizon is now the nation's largest wireless company, with 144 million customers and $74 billion in wireless revenue in 2022,[2] but only several hundred Verizon Wireless employees have union representation. Verizon Wireless persists in its anti-union activities. Yet, courageous Verizon workers continue to reach out to CWA for representation, winning union election at several retail stores in New York City in 2014, where they joined a unit of New York City Verizon Wireless technicians first organized in 1989. More recently, Verizon Wireless workers have joined the post-COVID upsurge of retail worker union organizing and won NLRB union elections and negotiated contracts in three locations (Seattle; Portland, Oregon; and Oswego, Illinois), despite continued aggressive anti-union activity by company management.[3]

Union density at Verizon Communications has declined to 22 percent. While the CWA wireline bargaining units at Verizon remain strong, negotiating good contracts and winning a militant six-week strike in 2016, they are an increasingly small island within a larger sea of 90,000 non-union employees. With the decline in the wireline business and because of outsourcing strategies, CWA today represents only about 3,900 call center workers at Verizon.[4]

As for AT&T, the study ends as CWA call center leaders were fighting a losing battle against the outsourcing of their work as the once-mighty corporation prepared itself for sale to SBC. Union power waned at AT&T, as the company lost market share to non-union companies and, aided and abetted by weak labor laws, succeeded in keeping the union out of its new lines of business. After SBC bought AT&T, the merged company (taking the AT&T name) became the second-largest private-sector union company in the nation (behind UPS), with about 68,000 union-represented employees in 2022.[5] In contrast to Verizon, AT&T respected the card check/neutrality agreements first negotiated with SBC and subsequently extended coverage to the companies it acquired. However, today the AT&T call center jobs are under enormous pressure as the company expands on the global outsourcing whose origins I trace in this study. In the past decade, AT&T closed forty-four in-house union centers, eliminated more than 16,000 call center jobs, and sent the work to third-party vendors in the United States and in the Philippines, India, Jamaica, Costa Rica, Dominican Republic, El Salvador, Columbia, and Canada. Base wage rates for contractors in the Philippines, Dominican Republic, and El Salvador ranged from $1.60 to $3.41 an hour in 2017, no match for the union rates that ranged from $25 to $40 an hour that year. Because of AT&T outsourcing and promotion of self-service strategies, the number of CWA-represented customer service employees at AT&T is about 15,000.[6] More recently, AT&T (and Verizon) have implemented so-called artificial intelligence (AI) software in portions of the customer service operations designed to further surveille and replace call center work.[7]

Before the Bell system monopoly's breakup, unions represented 63 percent of the telecommunications workforce. The unions effectively bargained for economic and working conditions for the entire sector. As the industry expanded, so did the union. The opposite is true today. Telecommunications is a dynamic industry, but growth is in the predominantly non-union wireless and broadband sectors. The non-union companies (the biggest of which are Comcast, Spectrum, Verizon Wireless, and T-Mobile) have fought aggressively to remain union free.[8] Today, union density in telecommunications hovers at just under 10 percent. Union wages, benefits, and working conditions no longer set the standard for the industry. Rather, non-union companies and global outsourcers exert downward pressure on union standards and employment security. The result is stagnant wages for telecommunications workers despite dramatic increases in productivity. Median telecommunications workers' wages barely changed over the almost forty-year period since divestiture, despite a 214-percent increase in productivity over the same period. Whether intended or not, neoliberal policymakers' promotion of competitive markets in a period of employers' aggressive anti-union animus is a direct cause.[9]

In recent years, CWA has recognized that it must adopt a global solidarity strategy and join with unions and worker centers seeking to organize call center workers (and others) in the Philippines, Dominican Republic, and Mexico. When CWA members went on strike against Verizon in 2016, Filipino workers at vendor call centers who were taking the re-routed Verizon calls reached out to CWA to offer solidarity. Later, CWA helped these workers fight their management's anti-union policies. Similarly, when an AT&T contractor in the Dominican Republic fired workers who were organizing a union, CWA pressed AT&T to make its contractor live up to AT&T's workers' rights policy. This helped achieve a historic agreement on organizing rights for Dominican workers at call center giant Teleperformance.[10]

The degradation and surveillance of work enabled by digital technologies that I describe in this book has become commonplace in workplaces across the United States, giving employers unparalleled ability to measure, control, speed up, degrade, and outsource work. Resistance to these practices has generated an upsurge of union organizing and militance in recent years. Amazon workers are organizing not only for better compensation and work schedules, but also to gain greater control over the digital tracking that follows warehouse workers every second on the job, releasing automated electronic messages in real time for excessive "time off task" subject to frequent discipline and dismissal. Amazon delivery drivers have rebelled against productivity benchmarks that eliminate downtime even for a bathroom break, as they post online photos of their urine-filled bottles (and force the company to admit in a letter to members of Congress that this does indeed take place). Railroad employees voted to go on strike in the fall of 2022, demanding, among other items, contractually protected sick days. Their employers' automated "precision scheduled railroading" software cut crews to the bone, eliminating any flexibility in the system. Hotel unions are looking for collective bargaining protections against just-in-time scheduling apps that send maids rushing with heavy carts up and down floors, disrupting their normal work rhythms. With the increase in at-home work during and after the COVID pandemic, some employers have installed surveillance equipment in their computer networks to calculate employees' "activity level." Software companies market monitoring software that can "view remote screens, track internet browsing history, record users' log in, log off, idle, and work time, view employees' screens in real time, encrypt this data, and prepare usage history logs and reports."[11]

In recent years, we have seen a flurry of organizing activity by non-union workers across the United States, including many in the service and retail sectors, from Starbucks baristas to Amazon warehouse stock pickers, from online and print journalists to bookstore employees. Workers in the technology sector, despite their reputed libertarian ethos and fiercely anti-union management, have been part of this movement, reaching out to CWA for assistance. Quality assurance workers at Activision Blizzard and other video game companies have selected union representation in NLRB elections. Retail workers at an Apple store in Oklahoma City voted for CWA. Tech workers at the *New York Times* formed the largest bargaining unit of technology employees in the United States with the News Guild/CWA. Google employees have formed a dues-paying Alphabet Workers Union, which is part of the CWA Campaign to Organize Digital Employees (CODE). Workers at Cognizant, an Alphabet subsidiary providing You Tube's music content operations, selected CWA representation. In this case, the NLRB affirmed that Cognizant and Google are joint employers, an important historic precedent. Gamers at SEGA of America have formed the Allied Employees Guild Improving SEGA. Workers at TCG Player, an eBay subsidiary, voted in the union in March 2023. Wireless tower technicians have come together to form Tower Climbers Union-CWA. It is too early to tell whether these initial efforts will result in negotiated contracts and achieve the scale and scope necessary to make lasting change.[12]

Most significant, as part of Microsoft's effort to gain regulatory approval of its $68 billion purchase of Activision, the company negotiated a groundbreaking labor neutrality agreement with CWA. Although the agreement would apply only to Activision workers following the transaction, Microsoft honored the same terms of non-interference and voluntary recognition when workers at its ZeniMax gaming subsidiary selected CWA representation in January 2023. General Electric signed a neutrality agreement with CWA should it win contracts to build two offshore wind turbine factories in New York State.[13]

Major breakthroughs for working people require a fundamental strengthening of our nation's labor laws and government regulation of corporate economic activity to shift the capital/labor balance. When Congress passed the National Labor Relations Act (NLRA) in 1935, it recognized the "inequality of bargaining power" between individual workers and employers.[14] The NLRA was designed to provide working people a collective voice through union representation to serve as a countervailing source of power to the wealth and influence of corporations. But the NLRA, designed for vertically integrated firms with relatively stable employment, leaves many gaps in today's political economy. Numerous proposals for change have surfaced, from strengthening the NLRA framework to more fundamental reform creating a U.S. version of European-style sectoral bargaining and works councils with broader authority over technology deployment and investment decisions. But even the modest Protecting the Right to Organize (PRO Act, H.R. 842), which passed in the Democratic-controlled U.S. House of Representatives in 2021, died when it failed to overcome a Senate filibuster.[15]

We have entered a new Gilded Age, one in which the fifty richest Americans hold more wealth than the combined total of the bottom half (165 million) of all Americans. Workers' wages have stagnated over the past four decades. Income inequality is now the highest it has been since the Census Bureau began tracking income distribution. With union representation in the private sector just below 7 percent, there is little, if any, countervailing force to counteract concentrated wealth and corporate power.[16]

The roots of the inequities in our society and economy lie in the neoliberal economic, trade, and labor policies that have given companies unfettered freedom to move jobs at will, depressing labor standards at home and abroad. The solutions to protect workers in this environment require government policies that strengthen organized worker power and social democratic regulation to constrain the destructive power of financial capital. To restore a modicum of balance for working people, we need fundamental reform of our labor laws, stronger financial regulation, strengthening of the social safety net, tax and budget policies that finance investments not only in our physical infrastructure but also, most important, in our people, and a restoration of our democratic institutions. Working people cannot be left on their own to battle the tremendous power of unrestrained capital; they need the support of government institutions to give them a fighting chance to come together to exercise collective power, to gain their fair share of the wealth that they create, and to win a measure of control over the great power of digital technology to improve—rather than further immiserate—conditions in our modern automated workplaces.

Appendix

Table A1. AT&T Union and Non-Union Employment, 1984–2004

	Total Employment	Union Employment	Non-Union Employment	Union Density
1984	365,000	226,300	138,700	62%
1985	338,000	201,110	136,890	60%
1986	317,000	183,860	133,140	58%
1987	303,000	169,680	133,320	56%
1988	304,500	158,340	146,160	52%
1989	283,500	147,420	136,080	52%
1990	274,000	128,780	145,220	47%
1991	317,100	123,669	193,431	39%
1992	312,700	121,953	190,747	39%
1993	308,700	108,045	200,655	35%
1994	304,500	106,575	197,925	35%
1995	300,000	76,000	224,000	25%
1996	130,400	50,840	79,560	39%
1997	128,000	58,560	69,440	46%
1998	107,800	42,000	65,800	39%
1999	148,000	39,990	108,010	27%
2000	166,000	35,424	130,576	21%
2001	117,800	31,806	85,994	27%
2002	71,000	26,270	44,730	37%
2003	61,600	22,176	39,424	36%
2004	47,600	16,660	30,940	35%

Source: AT&T Form 10-K and Annual Reports, various years

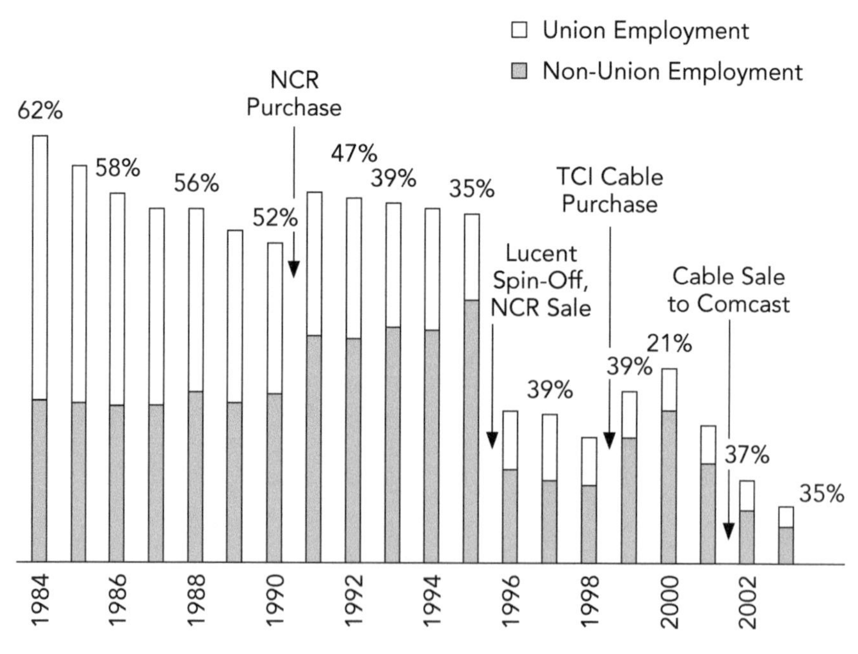

AT&T Union Employment as Percent of Total Employment, 1984–2004.

Table A2. Bell Atlantic Union and Non-Union Employment

	Total Employment	Union Employment	Non-Union Employment	Union Density
1984	82,800	53,820	28,980	65%
1987	80,950	51,808	29,142	64%
1988	81,000	51,840	29,160	64%
1990	81,600	52,224	29,376	64%
1991	75,700	51,476	24,224	68%
1992	71,400	47,124	24,276	66%
1993	73,600	47,840	25,760	65%
1994	72,300	46,995	25,305	65%
1995	62,600	43,820	18,780	70%
1996	141,000	98,700	42,300	70%
1997	140,000	96,600	43,400	69%
1999	145,000	100,050	44,950	69%
2000	260,000	137,800	122,200	53%
2002	229,500	112,455	117,045	49%
2003	203,100	103,581	99,519	51%
2004	210,000	102,900	107,100	49%

Source: Bell Atlantic SEC Form 10-K, various years

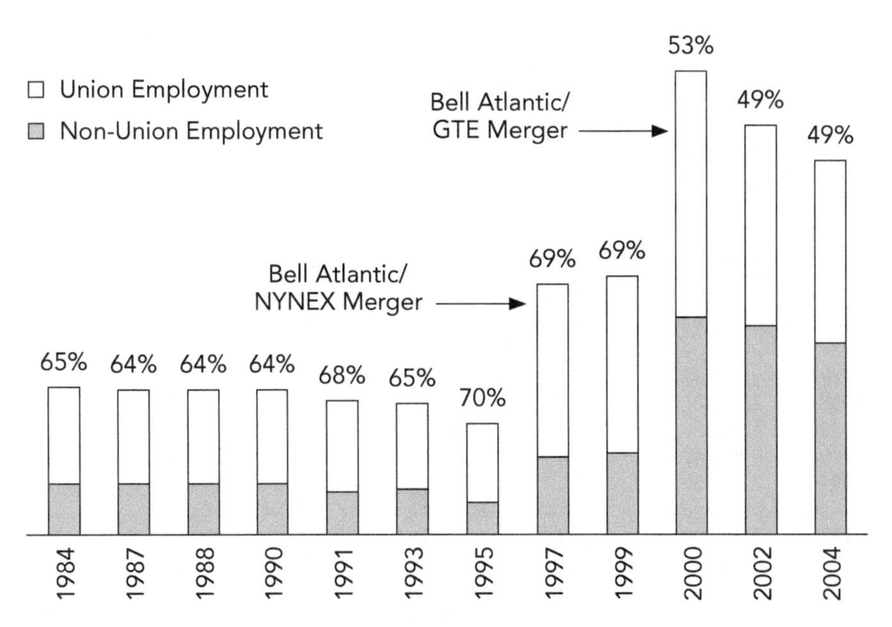

Bell Atlantic Union Employment as Percent of Total Employment, 1984–2004.

Table A3. AT&T Financial Performance, 1984–2004

	Revenue ($ billions)	Operating Income ($ billions)	Net Income ($ billions)	Operating Income/ Revenue	Return on Common Equity
1984	60,326	2,825	1,712	4.7%	9.8%
1985	63,159	3,562	1,856	5.6%	10.1%
1986	61,975	978	434	1.6%	0.3%
1987	60,726	4,164	2,374	6.9%	14.4%
1988	62,067	(2,381)	(1,527)	–3.8%	–11.3%
1989	61,604	4,931	2,820	8.0%	22.2%
1990	63,228	5,622	3,666	8.9%	19.7%
1991	64,455	1,570	171	2.4%	3.1%
1992	66,647	6,628	3,442	10.0%	21.1%
1993	69,351	6,568	(5,906)	9.5%	–29.0%
1994	75,094	8,030	4,710	10.7%	29.5%
1995	79,609	1,215	139	1.5%	0.7%
1996	52,184	8,810	5,908	16.9%	28.0%
1997	51,319	6,968	4,638	13.3%	21.5%
1998	53,223	7,487	6,398	14.1%	25.3%
1999	62,391	10,859	3,428	17.4%	15.2%
2000	65,981	4,277	4,669	6.5%	27.3%
2001	52,550	3,754	(4,131)	7.1%	22.1%
2002	37,827	4,361	963	11.5%	22.3%
2003	34,529	3,657	1,863	10.6%	16.4%
2004	30,537	–10,088	–6,469	–33.0%	12.6%

Table A4. Bell Atlantic Financial Performance, 1984–2004

	Revenue ($ billions)	Operating Income ($ billions)	Net Income ($ billions)	Operating Income/ Revenue	Return on Common Equity	Access Lines (000s)
1984	8,096	2,107	973.1	26%	13.3%	14,677
1985	9,131	2,325	1,093	25%	14.1%	15,090
1986	10,054	2,476	1,167	25%	14.3%	15,509
1987	10,747	2,383	1,240	22%	14.4%	16,056
1988	10,880	2,446	1,317	22%	14.5%	16,541
1989	11,595	2,008	1,024	17%	N/A	17,056
1990	12,547	2,614	1,231	21%	14.4%	17,484
1991	12,552	2,525	-324	20%	-4.4%	17,750
1992	12,718	2,506	1,341	20%	17.4%	18,181
1993	12,990	2,798	1,403	22%	17.3%	18,645
1994	13,791	2,805	-755	20%	14.4%	19,168
1995	13,081	2,937	1,882	22%	25.7%	20,566
1996	30,194	5,342	2,455	18%	19.1%	N/A
1997	31,566	6,627	4,202	21%	41.0%	41,600
1999	33,174	8,495	4,202	26%	41.2%	43,000
2000	64,707	16,758	11,797	26%	35.4%	64,900
2001	67,190	11,532	389	17%	1.7%	123,100
2002	67,625	14,997	4,079	22%	14.1%	128,137
2003	67,468	7,407	3,077	11%	10.5%	123,012
2004	71,283	13,117	7,831	18%	16.6%	116,861

Table A5. Timeline of Telecommunications Industry and CWA Milestones

1934	Communications Act of 1934
1947	CWA founded from Federation of National Telephone Workers
1953	CWA strike against BellSouth
1968	FCC approves Carterfone attachment to the Bell network
1968	CWA strike against the Bell system
1971	CWA strike against the Bell system
1969	FCC grants MCI application for Chicago–St. Louis microwave service
1971	FCC Ozark Plan increases cross subsidies from long-distance to local service
1973	EEOC Affirmative Action Consent Decree with AT&T
1981–1984	FCC grants two cellular licenses per region, one to local RBOCs
1982	AT&T Divestiture Consent Decree
1983	CWA strike against Bell system
1984	Divestiture takes effect
1986	CWA strike against AT&T
1989	CWA strike against NYNEX
1991	AT&T purchases NCR computer company
1993	FCC auctions PCS broadband wireless licenses
1994	AT&T purchases McCaw wireless
1996	Telecommunications Act of 1996
1996	AT&T spins off Lucent Technologies, sells NCR computer company
1997	Bell Atlantic/NYNEX merger
1997	SBC/Pacific Telesis merger
1998	AT&T purchases TCI cable company
1998	AT&T purchases Teleport
1998	CWA strike against Bell Atlantic
1999	SBC/Ameritech merger
1999	Bell Atlantic approved for long-distance service in New York State
2000	Bell Atlantic/GTE merger to form Verizon
2000	CWA strike against Bell Atlantic
2001	AT&T spins off wireless business
2002	AT&T sells cable business to Comcast
2004	FCC *UNE Remand* decision ends below-cost local service resale pricing
2004	AT&T exits local service business
2005	SBC purchases AT&T (takes AT&T name)
2006	Verizon purchases MCI

Notes

Introduction

1. Karasek, "Job Demands, Job Decision Latitude, and Mental Strain."

2. Karl Marx, *The Eighteenth Brumaire of Louis Bonaparte* in *Karl Marx: A Reader*, ed. John Elster. For a discussion of the struggles between capital and labor over technology deployment, see Noble, *Forces of Production*; Burroway, *Manufacturing Consent*; Braverman, *Labor and Monopoly Capital*; Resnikoff, *Labor's End.*

3. For discussion of worker agency in highly automated call centers, see Taylor and Bain, "An Assembly-Line in the Head"; Callaghan and Thompson, "Edwards Revisited." For the union role in call centers, see Doellgast and O'Brady, "Collective Voice and Worker Well-Being." For the argument of total management control in call centers, see Fernie and Metcalf, *(Not) Hanging on the Telephone.*

4. For discussion of CWA operators, see Norwood, *Labor's Flaming Youth,* and Green, *Race on the Line*. For a discussion of the power of autonomous female craft unions, see Cobble, *Dishing It Out*, 1–14. For a discussion of the challenges women workers faced in amalgamated industrial unions, see Gabin, *Feminism in the Labor Movement.*

5. Hochschild, *The Managed Heart.* Among the now extensive literature on unionized women service workers, see Cobble, *Dishing It Out*; Benson, *Counter Cultures*; Nielsen, *From Sky Girl to Flight Attendant*; Barry, *Femininity in Flight*; Murphy, *Deregulating Desire*; Boris and Klein, *Caring for America;* Cassedy, *Working 9 to 5*; Parker, *Department Stores and the Black Freedom Movement*; Deslippe, *Rights, Not Roses*; Kessler-Harris, *Gendering Labor History*.

6. For a discussion of the 1970s as "the last days of the working class," see Jefferson Cowie, *Stayin' Alive*. For scholarship that complicates this view by bringing women and people of color into the narrative, see Lane Windham's magisterial *Knocking at Labor's Door*; Hower, "You've Come a Long

Way—Maybe" (on public sector comparable worth campaigns); Turk, *Equality on Trial* (on the fight for comparable worth); Murphy, *Deregulating Desire*; Barry, *Femininity in Flight* (on flight attendants); Nadasen, *Household Workers Unite*; Cobble, "A 'Tiger by the Toenail'" and "A Spontaneous Loss of Enthusiasm." For a broader discussion on labor feminists' fight for equity and economic justice at the workplace and in social policy, see MacLean, *Freedom Is Not Enough*; Cobble, *The Other Women's Movement*; Kessler-Harris, *In Pursuit of Equity*; Milkman, *A Century of U.S. Women's Labor History*.

7. Turk, *Equality on Trial*, 8.

8. Historian Nelson Lichtenstein, in his authoritative *State of the Union* survey of the twentieth-century union movement, concludes that an individualistic rights consciousness, flowing from passage of the Civil Rights Act of 1965, has eclipsed a more solidaristic labor rights framework focused on strengthening strong unions. An individualistic rights-based approach does not fundamentally shift the power of capital. While he acknowledges that collective action, institution building, and rights consciousness are not mutually exclusive, he argues that the goals of equity and economic justice can best be achieved through democratic unions. Lichtenstein, *State of the Union*, 178–211. Other historians emphasize the connection between civil rights, women's rights, and labor rights. See Windham, *Knocking at Labor's Door*; Turk, *Equality on Trial*; Parker, *Department Stores and the Black Freedom Movement*; Hower, "You've Come a Long Way—Maybe."

9. On the rise of the consumer economy, see Cohen, *A Consumers' Republic*. On the growth of call centers, see Stevens, *Call Centers and the Global Division of Labor*, 2.

10. The source for the 3.7 million statistic is ContactBabel, *US Contact Centers 2023–27* for the year 2022. Total non-farm employment in 2022 was 223 million. U.S. Bureau of Labor Statistics, Current Employment Survey, Dec. 2022 (3.7 million divided by 223 million = 2 percent). The U.S. Office of Management and Budget's North American Industry Classification System (NAICS) does not have a code for the "call center industry." The U.S. Bureau of Labor Statistics counts 2.9 million customer service representatives as of May 2022. BLS, Occupational employment and wages, May 2022, for occupational code 43-4051.

11. The number of production workers in December 2022 for machinery manufacturing (NAICS code 333) was 1.1 million; for transportation equipment manufacturing (NAICS code 336), 1.7 million; for primary metal manufacturing (NAICS code 331), 366,000, for a total of 3.1 million. U.S. Bureau of Labor Statistics, Current Employment Survey, Table B-1a (Establishment data) for December 2022.

12. There is no data source on the size of the global call center industry. Country estimates in the early 2000s include United Kingdom (800,000), Canada (500,000), Germany (400,000), Austria (250,000), India (500,000), and South Africa (100,000). Holman, Batt, and Holtgrewe, *The Global Call*

Center Report. The Philippines has now eclipsed India as the leading location of offshore call centers; in 2012, scholar Jan Maghinay Padios reported 400,000 Filipino call center workers, and in 2015, journalist accounts cited 1 million Filipino call center workers. Padios, "Listening Between the Lines," 3; Don Lee, "The Philippines Has Become the Call Center Capital of the World," *Los Angeles Times*, February 1, 2015.

13. StrategyR, Global Industry Analysts, Inc. https://www.strategyr.com/market-report-call-centers-forecasts-global-industry-analysts-inc.asp. The four global call center companies are Concentrix (which merged with Convergys in 2018), Teleperformance, TTEC Holdings, and Sitel (which merged with Sykes in 2021 and rebranded itself as Foundever in 2023). Concentrix SEC Form 10-K for the year ended November 30, 2022; TTEC Holdings SEC Form 10-K for the year ended December 31, 2022; Sitel/Foundever website visited June 8, 2023; Teleperformance 2022 Universal Registration Document, available on the company website at http://teleperformance.com.

14. There is extensive literature on the Bell system breakup. A good start is Temin and Galambos, *Fall of the Bell System*, 269–276; see also chapter 2.

15. In December 2021, Bell system retirees posted on Facebook their views of the impact of divestiture on customer service. Here is a sampling of their comments: "38 years ago tonight we started a journey that supposedly spawned the information age. There were a lot of casualties along the way myself included. When I see what is left of the company, the outside plant and most importantly the pride in service I feel a deep sadness of what once was!" "It is truly sad to witness the changes, especially the lack of customer service. Thankful we feel like we made a difference in customer service while we were working." "We were a monopoly and WE CARED! Competition was supposed to improve that . . . you be the judge!!!" "Unfortunately 'service' hasn't been part of the corporate cost model since after divestiture. It's all about return on investment and keeping stock holders happy. That's been a struggle as well." "It was an unfortunate event that Judge Green destroyed something that was working so well. A lot of employees were left as roadkill." "When there was no competition, our customers and employes were treated like gold. After competition became reality, the customer was no longer first and definitely not the employees. Judging by the results, not a great business model." "I remember when Charles Brown [CEO of AT&T at the time of divestiture] said we were a perfectly balanced stool with 3 legs. 1 leg was the customer, 1 was the stockholder and 1 was the employee. I wouldn't have wanted to sit on that stool by the time I retired." "The best service in the world was destroyed by a single stroke of a pen." "I started in 1968 and retired in 2000. My biggest frustration today is not being able to reach a customer service representative when calling businesses, without going through all of the prompts. I remember we had to answer customer calls within a designated time or the PSC was involved. Those were the good days." "As most of you will remember, we had 24 hours to restore an out of service customer from the time we received the

trouble ticket. It is upwards of 2 weeks in today's world. So sad." Facebook posts dated December 31, 2021, in author's possession.

16. Schmitt and Kandra, *Decades of Slow Wage Growth for Telecommunications Workers,* Table 4, 8 (for 1980s); Hirsch and MacPherson, "Union Membership, Coverage, Density and Employment from the CPS," http://unionstats.com (for 2005); Bureau of Labor Statistics, "Union Members-2023," Table 3, January 23, 2024 (for 2023).

17. Union representation across the entire economy (including both private and public sectors) was 11.3 percent in 2022. Bureau of Labor Statistics, "Union Members 2022," Tables 1 and 3, January 19, 2023.

18. Doellgast, *Exit, Voice, and Solidarity,* 50–104. This study of labor relations in telecommunications companies in ten countries finds that call center workers have greater control over conditions at work and job security in Scandinavia, Germany, and Austria than in the liberal market economies of the United States and Great Britain. However, in recent years, increased outsourcing by Deutsche Telekom, Germany's largest telecommunications company, has led the telecom union ver.di to trade concessions over work rules for greater job security.

19. Citation from Doellgast, *Disintegrating Democracy at Work,* ix–xx; See also Doellgast, *Exit, Voice, and Solidarity*; Holman, Batt, and Holtgrewe, *The Global Call Center Report,* vi–vii; Batt, Holman, and Holtgrewe, "The Globalization of Service Work."

20. Colin Lecher, "How Amazon Automatically Tracks and Fires Workers for 'Productivity,'" *The Verge,* April 25, 2019; Jodi Kantor, Karen Weise, and Grace Ashford, "Power and Peril: Five Takeaways on Amazon's Employment Machine," *The New York Times,* June 15, 2021; Dan Clawson and Mary Ann Clawson, "IT is Watching: Surveillance and Worker Resistance," *New Labor Forum* 26, no. 2 (2017): 62–69; Zickuhr, *Workplace Surveillance is Becoming the New Normal for Workers*; Mateescu and Nguyen, "Algorithmic Management in the Workplace"; Levy, *Data Driven*; Jodi Kantor and Arya Sundarm, "The Rise of the Worker Productivity Score," *The New York Times,* August 14, 2022; Danielle Abril and Drew Harwell, "Keystroke Tracking, Screenshots, and Facial Recognition: The Boss May Be Watching Long after the Pandemic Ends," *The Washington Post,* September 24, 2021.

21. Taylor, *The Principles of Scientific Management*; Zuboff, *The Age of Surveillance Capitalism.*

22. Montgomery, *Workers' Control in America,* 154. Montgomery's scholarship focused on workers' control during and after the transition from craft to industrial employment.

23. Important studies of workers' struggles over job control include Montgomery, *Workers' Control in America*; Edwards, *Contested Terrain*; Meyer, *The Five Dollar Day*; Metzgar, *Striking Steel,* 94–117; Moberg, "Rattling the Golden Chains: Conflict and Consciousness of Auto Workers," (on Lordstown); Davies,

Women's Place Is at the Typewriter; Benson, *Counter Cultures*; Green, *Race on the Line*; Cobble, *Dishing It Out*.

24. This framework comes from Richard Edwards, *Contested Terrain,* 15.

25. Historian Gary Gerstle defines a political order as a "constellation of ideologies, policies, and constituencies that shape American politics in ways that endure beyond the two-, four-, and six-year election cycles." Gerstle's chronology locates the beginnings of the Neoliberal Order in the Carter administration in the late 1970s through its heyday in the 1980s during the Reagan administration, continuing through the Clinton and Bush administrations, and falling apart with the Obama administration's response to the 2008 recession and the ascendance of both Donald Trump and Bernie Sanders in the 2020 campaign. In contrast, Lily Geismer distinguishes between the neoliberalism of the Clintonian New Democrats and the more extreme market orientation of Republican neoliberals. Fraser and Gerstle, *The Rise and Fall of the New Deal Order,* and for a reevaluation, Gerstle, Lichtenstein, and O'Connor, *Beyond the New Deal Order*; Geismer, *Left Behind,* 7–13; Gerstle, *The Rise and Fall of the Neoliberal Order*, 2. On the rise of neoliberalism, a good place to start is Kotz, *The Rise and Fall of Neoliberal Capitalism.*

26. The literature on financialization is quite extensive. Major works include Krippner, *Capitalizing on Crisis*; Lazonick, "Profits without Prosperity"; Davis, *Managed by the Markets*; Appelbaum and Batt, *Private Equity at Work*; Jacoby, *Labor in the Age of Finance.*

27. Friedman, "The Social Responsibility of Business Is to Increase Its Profits," 17.

28. Monopoly-era AT&T was the poster child of the Chandlerian corporation with the "visible hand" of management vertically integrating all company functions within the firm for maximum efficiency. Chandler, *The Visible Hand,* especially 195–206, for discussion of AT&T organizational structure. The classic economic argument for the efficiencies of vertical integration is Coase, "The Nature of the Firm." On the link between financialization and the fissured workplace, see Weil, *The Fissured Workplace,* 8–1. For a good description of the financial turn in management, see Batt, "The Financial Model of the Firm, the 'Future of Work,' and Employment Relations."

29. There is extensive literature on the failure of labor law to protect organizing workers. A good place to start is Freeman and Medoff, *What Do Unions Do?* and Freeman's update in "What Do Unions Do: The 2004 M-Brane Stringtwister Edition," in *What Do Unions Do? A Twenty-Year Perspective,* eds. Bennett and Kaufman, 627–629; Weiler, *Governing the Workplace,* 112, 238–239, and n. 18; Windham, *Knocking on Labor's Door*; McNicholas et al., "Unlawful: U.S. Employers Are Charged with Violating Law in 41.5% of All Union Election Campaigns." For management accounts of employer tactics to defeat union organizing efforts, see Levitt and Conrow, *Confessions of a Union Buster* and Demaria, *How Management Wins Union Organizing Campaigns.*

Chapter 1. Before the Breakup

1. Kmetyk, interviews with author.

2. Fraser and Gerstle, *The Rise and Fall of the New Deal Order*; for a reevaluation, see Gerstle, Lichtenstein, and O'Connor, *Beyond the New Deal Order*; on managerial capitalism, see Chandler, *Visible Hand*, especially 200–203 on the Bell system.

3. Kleinfeld, *The Biggest Company on Earth*, 6; Temin and Galambos, *Fall of the Bell System*, 11–12; *In the Matter of Petitions Filed by the Equal Employment Opportunity Commission (EEOC)* et al., Memorandum Accompanying the August 1, 1972, Submission of the Bell Companies, FCC Records, Docketed Case Files 1927–90, Docket No. 19143, Record Group 173, National Archives at College Park, MD, box 565, vol. 12, 22–24 (hereafter EEOC case). At some point, the National Archives renumbered the boxes containing the AT&T/EEOC case files. All citations to the EEOC case materials are to the renumbered boxes.

4. Citation from 1907 AT&T Annual Report in Garnet, *The Telephone Enterprise*; see also Schlesinger et al., *Chronicles of Corporate Change*, 11; John, *Network Nation*.

5. Quotation in Schlesinger et al., *Chronicles of Corporate Change*, 12–13. See also Temin and Galambos, *Fall of the Bell System*, 23–27, 54, 59, 126.

6. Author calculations based on Federal Communications Commission, *Statistics of Communications Common Carriers*, Table 10, 1969 (for the year ending December 31, 1967) and 1983 (for the year ending December 31, 1981); See also *Bell Statistical Manual*, 1980, 702; Kleinfeld, *The Biggest Company on Earth*.

7. Testimony of Dr. Ann Scott, EEOC Hearing, Washington, D.C., June 5, 1972, EEOC case, box 565, vol. 11, page 4972 of hearing transcript.

8. Green, *Race on the Line*, 53–55.

9. Author calculation based on EEOC statistical database, EEOC case, exhibit 6, box 160, vol. 26. Service representatives were a subset of customer service employees.

10. Author calculations from FCC, *Statistics of Communications Common Carriers*, Table 10, 1969 and 1983; See also Kenneth Lipartito, "When Women Were Switches."

11. EEOC Findings of Fact, Summary of Exhibits, January 31, 1972, EEOC case, box 567, vol. 21 for Exhibits C-1111, C-1112, C-216, R-1111, Z-157, Z-172, Z-698; vol. 22 for Exhibit R-1174.

12. Letter from John W. Kingsbury, AT&T Assistant Vice-President, to All Personnel Vice Presidents and to All AT&T Assistant Vice-Presidents, August 12, 1971, EEOC case, box 577, vol. 52.

13. Walter W. Straley, "Force Loss and the Urban Market," *Report to Bell System Presidents' Conference*, October 9, 1969, EEOC Exhibit 5, EEOC case, box 568, vol. 23; author calculations based on EEOC statistical database. EEOC case, box 568, vol. 6.

14. Salaam and Hargrove, interviews with author. Ameenah Salaam worked

for seven years as a service representative for Diamond State Telephone in Delaware in the 1990s before promotion to CWA staff representative. She then rose to become assistant to the vice president in California, assistant to CWA President Chris Shelton in 2017, and secretary-treasurer of the national union in 2023. Elizabeth Hargrove worked for 30 years for C&P Telephone as an operator and later service representative, 1970–2000.

15. Author calculation based on EEOC statistical database, EEOC Exhibit 6, EEOC case, box 568, vol. 26.

16. AT&T SEC Form 10K for the year ended December 31, 1982 (for total employment and union employment); FCC, *Statistics of Common Carriers*, Table 10, 1983 (for the number of non-management employees). Sandy Kmetyk, whom we met at the beginning of this chapter, became a CWA member when the Pennsylvania Telephone Guild merged into CWA after AT&T divestiture in 1985. Dina Beaumont, whom we meet later in this chapter, became a CWA member when the Federation of Women Telephone Workers joined CWA in 1974.

17. Norwood, *Labor's Flaming Youth*, 1–24, 169–198; Green, *Race on the Line*, 89–114; Brooks, *Communications Workers of America*, 14–16, 19–26.

18. For example, Bell of Pennsylvania in 1940 voluntarily recognized the former company union as the independent Pennsylvania Telephone Guild. See Agreement between Commercial Department Unit of the Plan of Employee Representation—Employees of the Bell Telephone Company of Pennsylvania and the Bell Telephone Company of Pennsylvania, March 11, 1938; Letter from Washington L. Hudson, President, Pennsylvania Telephone Guild, to Mr. J.T. Harris, Vice President, Bell Telephone Company of Pennsylvania, March 5, 1940; Letter from John T. Harris to Washington L. Hudson, President, Pennsylvania Telephone Guild, March 8, 1940; Letter from Thomas H. Griest, General Commercial Manager, Bell Telephone Company of Pennsylvania, August 7, 1940. All documents in authors' possession.

19. Brooks, *Communications Workers of America*; CWA, *CWA at Fifty*; Joseph Beirne Testimony, Hearings before the Subcommittee on Labor-Management Relations of the Committee on Labor and Public Welfare, United States Senate, Eighty-first Congress, Second Session, August 10, 1950; Schacht, "Toward Industrial Unionism."

20. Beaumont, interview with author. In 1974, 15 percent of CWA local presidents were women, although more than half of all secretary-treasurers and secretaries were women. CWA memo, "Breakdown by Sex of Local Officers," February 1974, The Communications Workers of America Records; Wagner 124; box 20, folder 9; Tamiment Library/Robert F. Wagner Labor Archives New York University (hereafter CWA-TL).

21. The two non-white men on the CWA executive board during this period were Louis Knecht, CWA secretary-treasurer (1974–1985), who was Native American, and Rudy Mendoza, national director, who was Hispanic. Herrera, interview with author.

22. John C. Carroll, Memo to Glenn E. Watts, Goals and Timetables and Authorized Staff Positions as of October 1, 1974, box 20, folder 9, CWA-TL; John C. Carroll, Memo to Joseph A. Beirne, Report of the Female Structure Study Committee, January 8, 1974, box 20, folder 4, CWA-TL.

23. The list does not include other minority groups. Check-List: Black Local Officers-Executive Board Members, CWA executive board minutes, May 2, 1973, box 20, folder 1, CWA-TL.

24. Daily Proceedings and Reports, CWA 35th Annual Convention, June 18–22, 1973, 22, 34, 162–64, 270–74, 333.

25. Report of the Female Structure Study Committee; John C. Carroll, Memo to CWA President Joseph A. Beirne, January 8, 1974; Joseph A. Beirne, Memo to CWA executive board, January 18, 1974, box 20, folder 4, CWA-TL; Letter from CWA Black Members to CWA executive board, May 21, 1971, box 19, folder 18, CWA-TL; CWA executive board minutes, February 15, 1974, box 20, folder 4, CWA-TL; Herrera, interview with author.

26. CWA, Report to the Communications Workers of America Committee of the Future, "Minority Involvement and Participation within CWA," nd, box 57, folder 7, CWA-TL; Proceedings of CWA 69th Annual Convention, July 16–17, 2007, 52–62; CWA web page (cwa-union.org) for 2023 data, visited on July 20, 2023.

27. Lorena Weeks Testimony, EEOC case, box 568, vol. 25; see also Stockford, *The Bellwomen*, 111–12; Herr, *Women, Power, and AT&T: Winning Rights in the Workplace*, 56–7, 76, 45–46, 81–82.

28. Historian Venus Green sharply criticizes the EEOC for its failure to consider racial discrimination alongside gender bias in the AT&T suit. As a result, according to Green, many Black women at AT&T were harmed by the final settlement, which opened Black employees' operator jobs to men just as the company was cutting operator positions, while at the same time, the settlement undervalued the monetary damages due Black women. Green, "Flawed Remedies"; Green, *Race on the Line*, 237–42.

29. Stockford, *The Bellwomen*; MacLean, *Freedom Is Not Enough*, 107, 131–33; Herr, *Women, Power, and AT&T: Winning Rights in the Workplace*, 142–54.

30. CWA Motion to Intervene, *Equal Employment Opportunity Commission and United States of America and Communications Workers of America v. AT&T* et al., Civil Action 73–149, March 12, 1973; Affidavit of Richard W. Hacker, Assistant to CWA President Joseph Beirne; EEOC Opposition to CWA Motion to Intervene. Joseph A. Beirne, Letter to Robert D. Lilley, AT&T Executive Vice President, December 14, 1971. All documents in EEOC case, box 567, vol. 19.

31. Testimony of Helen J. Roig, EEOC case, box 568, vol. 25. Roig also filed a CWA Job Pressures survey into the record. EEOC case, box 568, vol. 25. Gay Semel, a New York Telephone Company operator, CWA organizer, and future CWA attorney, provided lengthy written and oral testimony at the New York hearing, May 8, 1972. Dennis Serrette, CWA Local 1101 officer in New York

City and leader of the National Black Communications Coalition, testified at the New York hearing, May 8, 1972. EEOC case, box 564, vol. 9.

32. This group includes Hazel Dellavia in New Jersey and Barbara (Lephardt) Fox Shiller in Maryland, for whom the consent decree opened the door to promotion from service representative to the previously all-male communications consultant position; Gail Evans in Maryland, who moved from service representative to a technician position and later became president of CWA Local 2100 for almost twenty years and eventually an administrative officer in CWA District 2/13; and Ron Collins in Maryland, who moved from his service representative position into local and then national union leadership. Annie Hill and Laura Unger were pioneer women technicians who rose through the ranks to high-level CWA positions responsible for bargaining and representation of call center workers. Annie Hill became vice president of CWA District 7 (representing fourteen western states) and CWA secretary-treasurer. Laura Unger was president of CWA Local 1150 in New York City and assistant to the vice president responsible for the AT&T bargaining unit.

33. AT&T Archives, File 549–05–01 (originally published in *Glamour, Mademoiselle*, June 1969). The ad was also filed in the EEOC case as EEOC Exhibit 8, Document No. C-471, EEOC case, box 568, vol. 25.

34. The classic discussion of gendered emotional labor is Hochschild, *The Managed Heart;* for a discussion of how flight attendants used their gender role to build unity and power, see Barry, *Femininity in Flight*, 2007.

35. Chesapeake and Potomac, Job Qualification Sheet for Service Representative Job Title, October 1975. CWA research department archives, box 635906, Job Titles—District 1 and 2 Notebook.

36. Hazel Dellavia, written response to author; Linda Kramer, Sandy Kmetyk, and Michele Guckert, interviews with author. Hazel Dellavia began work as a New Jersey Bell operator before promotion to a service representative and later communications consultant. She was president of CWA Local 1024 in New Jersey and then CWA staff representative. Linda Kramer was president of CWA Local 1023 in New Jersey, 1995–2004. Michele Guckert began working at C&P Telephone in 1973 as a clerk typist before her promotion to service representative in 1975 and later supervisor of a residential customer call center until her retirement in 2001.

37. Dellavia, written response to author; Collins, interview with author; Langer, "Inside the New York Telephone Company," 21. Langer, a college graduate and self-described radical, took a job with New York Tel to understand, in her words, why members of the "new—white collar—'working class' did not 'rise up' against their employers." Langer's description of the service representative job as one with little autonomy differs from the descriptions that service representatives I interviewed provided. It is likely that my sources, who worked for the telephone companies for decades, not just three months, developed a greater sense of autonomy over time, and they talked

about this period in contrast to the intensive job control after digital systems were introduced.

38. Kzirian, interview with author. Kzirian began work with New York Telephone in 1968 as a manager in a business customer service office. He was promoted to various management customer support positions at New York Telephone Company and moved over to AT&T at the 1984 divestiture as director of customer care for equipment and long-distance services and operations. From 1994 to 2004, he was AT&T telemarketing sales vice president responsible for a $500 million global operation in seventy-two call centers in eight countries with 15,000 agents; Dellavia, written response to author; Guckert and Collins, interviews with author; Elinor Langer, "Inside the New York Telephone Company."

39. Collins and Kramer, interviews with author; Dellavia, written response to author.

40. EEOC case, box 568, vol. 25.

41. Collins and Kramer, interviews with author; Dellavia; written response to author.

42. Collins and Kramer, interviews with author; Dellavia; written response to author. See chapter 6 for discussion of Verizon strike and the fight over "closed key" time off the phones.

43. Service representatives had additional miscellaneous job functions, including permanent disconnects ("put all customer records in 'Final Bill' envelope—form 3540"); preparing "Uncollectible Account Vouchers" (form 3732); preparing telephone directory orders (form 3650); and entering customer non-published telephone number agreements (form 1–3621–4). At the end of each day, the service representative was required to prepare a "Service Representative Contact and Work Item Record" from the stub of the "Contact Memorandum" (form E-2921). Analysis of Job Requirements: Office Occupations—Service Representative, Residence Contacts, prepared by Philadelphia Operations District Manager, May 14, 1965. EEOC case, box 169, vol. 61.

44. Langer, "Inside the New York Telephone Company," 21–22.

45. Langer, "Inside the New York Telephone Company," 21–22; Dellavia, written response to author; Guckert and Kmetyk, interviews with author; Testimony of Helen J. Roig, February 9, 1972, EEOC case, box 568, vol 25; EEOC Findings of Fact, EEOC case, box 567, vol. 21.

46. Stake, interview with author. Stake began his thirty-four-year career with AT&T Long Lines in 1970 as a traffic engineer, moved into management positions in sales and marketing in AT&T's general business and national accounts divisions servicing business customers, and after divestiture, served as AT&T vice president for business customer care and later vice president for sales and customer care in AT&T's consumer services division until his retirement in 2004. The source for the 1,500 service representative figure is Kansas City Local 6450 1980 Long Lines Bargaining Demand, presumably

1979 or 1980, box 903997, CWA communications and technologies department archives; 1980 Bargaining Demands, box 123, folder 123, CWA-TL.

47. Memo from Elaine T. Gleason, National Director to Joseph A. Beirne, CWA President re Commercial Representatives, August 7, 1957, box 123, folder 32, CWA-TL; CWA/AT&T contracts, Appendix 1, for the years 1948–1977, CWA communications and technologies department; CWA Bargaining Demand Submitted by Local 1152, 1980, box 12, CWA-TL; Proper Wage Alignment for Marketing/Sales Clericals, 1983 Bargaining, presumably 1983, box 90402, CWA communications and technologies department archives.

48. Dellavia, written response to author; Kramer, Kmetyk, Fox Shiller, and Evans, interviews with author. Fox Shiller (nee Lephardt) began work as a service representative for New York Telephone in Brooklyn in 1972, subsequently moved to Richmond, Virginia, where she took a job with C&P Telephone in an office serving business customers, and in 1977 was promoted to communications representative. In 1981, she left C&P to work for CWA, eventually becoming the most senior staff in CWA District 2 (Maryland, Virginia, West Virginia, and Washington D.C.) as assistant to vice president. She chaired CWA bargaining committees with Bell Atlantic in 1989, 1992, and 1995 and retired from CWA in 2003.

49. Dellavia, written response to author.

50. Dellavia, written response to author; Collins, interview with author; Langer, "Inside New York Telephone Company."

51. John Whetzell, "Absence Control in the Telephone Industry," January 4, 1974, 1974 Bell System Bargaining Notebook, CWA District 2; Absence Rates of Employees, EEOC case, box 165, vol. 4 (document filed with Bell of PA and Bell of Diamond State (DE) materials).

52. Walter Straley, AT&T Vice President, "Report on Force Loss and the Urban Labor Market," October 9, 1969, Document C-1540, EEOC Exhibit 5, EEOC case, box 569, vol. 23; Washington Commercial Personnel Survey, "The Service Representative: Her Story," 1970, EEOC case, box 568, vol. 26.

53. Bell Companies: All Employees by Department and Length-of-Service, Table 2A, Bell Company Exhibit #1, EEOC case, box 569, vol. 29.

54. Spalter-Roth and Hartmann, "Women in Telecommunications"; author calculations from CWA contracts with C&P, New Jersey Bell, AT&T Long Lines, and Bell of Pennsylvania, various years (for service representatives' annual earnings); Bureau of Labor Statistics, "Median Usual Weekly Earnings of Full-Time Wage and Salary Workers 25 Years and over by Sex and Educational Attainment, Annual Averages 1979–2013" and U.S. Census Bureau, Current Population Survey, Annual Social and Economic Supplements, Table P-38, "Full-Time, Year-Round All Workers by Median Earnings and Sex: 1960 to 2013" (for female and male workers' annual earnings).

55. In 1983, technicians at C&P earned about $30,000 a year, a 20-percent wage premium over the service representative earnings of about $24,000.

Operators' annual earnings of about $21,000 in 1983 trailed those of service representatives by about $2,500 a year (11 percent). The service representative/operator pay gap began to widen in the late 1970s, reflecting the growing importance of the service representative occupation. Author calculations based on CWA wage rates derived from CWA contracts with C&P, New Jersey Bell, AT&T Long Lines, and Bell of Pennsylvania, various years, and John E. Strouse, CWA development and research department, "Bell System Wage Rate Changes Since 1957: Top Rate and Minimum Rates Analyzed in CWA Districts for Selected Major Cities," 1974 Bell Bargaining notebook, CWA District 2 office.

56. Green, *Race on the Line*, 73–76, 46–48, 117–136, 215–19, 227; Vallas, *Power in the Workplace*, 95–100.

57. Goldman, "Curbing Big Brother in the Workplace."

58. Vallas, *Power in the Workplace*, 83–140.

59. None of these items appears on a CWA list of "Bell System bargaining council objectives, attained and unattained, 1958–69," that was produced in preparation for 1970 Bell bargaining. Bell System Bargaining Council Objectives (Attained and Unattained, 1958–1969), January 29, 1970, box 19, folder 1, CWA-TL.

60. Chronological Outline of Job Pressures Committee Action, attached to CWA executive board minutes, January 11, 1973, box 19, folder 16, CWA-TL.

61. The members of the Job Pressures Study Committee consisted of three women and three men: Chair Clara Allen, New Jersey Area Director; Robert Butland, Local 1022; McCoy Garrison, Local 2204; Faye Holub, Local 6312; Bertha Van Sittert, Local 6323; Ellis Crandell, Local 9409; and Myrtle Robertson, Local 3372. Chronological Outline of Job Pressures Committee Action, Attached to CWA executive board minutes, January 11, 1973, box 19, folder 16, CWA-TL; Final Job Pressures Report, June 4, 1969, box 146, folder 19, CWA-TL; Job Pressures Committee memo to President Joseph A. Beirne, October. 31, 1969. The memo is marked "Personal and Confidential." Box 19, folder 1, CWA-TL. See also CWA 32nd Annual Convention Proceedings, 1970, 142–44.

62. Clara Allen chaired the job pressures implementation committee. Clara Allen was a longtime leader in CWA District 1 in the Northeast. The other two members of the committee were Patsy Fryman, CWA staff representative from the Midwest, and Victor Crawley, CWA state director for Missouri, Arkansas, and Illinois. We will meet Crawley again in chapter 3 as vice president of CWA District 6, when he led the successful bargain-to-organize campaign at Southwestern Bell Mobile Systems. Final Job Pressures Report, June 4, 1969, box 146, folder 19, CWA-TL. See also Chronological Outline of Job Pressures Committee Action, box 19, folder 16, CWA-TL; CWA Executive Board minutes, June 1969, box 146, folder 27, CWA-TL; Memo from Clara Allen to Joseph A. Beirne, President, Job Pressures Implementation Report, September 18, 1970, box 19, folder 6, CWA-TL.

63. CWA progress in promoting women can be seen in the makeup of the 1978 national women's committee. Patsy Fryman was now assistant to CWA President Watts, Selina Burch was headquarters staff, and Maxine Lee, LaRene Paul, and Florine Poole and new member Barbara Easterling (who later became CWA secretary-treasurer) held administrative positions in their districts. Dina Beaumont was a CWA vice president. CWA National Women's Conference Program, September 28-October 1, 1978, box 21, folder 15, CWA-TL; Job Pressures Resolution, box 22, folder 1, CWA-TL; CWA Executive Board minutes, Box 22, folder 2, CWA-TL.

64. CWA Flyer, "Assembly Lines Are Only in Factories. Right? Wrong!," 1979, box 159, folder 1, CWA-TL.

65. Letter from Rex R. Reed to Glen Watts, August 9, 1980, box 109, folder 13, CWA-TL (for CWA/AT&T monitoring language); CWA/C&P Contract, August 10, 1980 (for BOC monitoring language); CWA development and research department, A Primer for Bell System Collective Bargaining in 1980, February 12, 1980, 272, box 124, folder 1, CWA-TL; Analysis of Bargaining Items Submitted to the Bell System Bargaining Council by Districts, Locals, Bargaining Units, and Individual Members, March 11, 1980, box 106, folder 1, CWA-TL; CWA's 1980 Bell System Bargaining Council Resolution, March 12, 1980, box 106, folder 1, CWA-TL (for bargaining demands).

66. CWA's focus on the impact of technological change on its members' job security and satisfaction was not new. As early as 1941, CWA's predecessor union, the National Federation of Telephone Workers, articulated a three-part response to technological change. First, the union recognized that, because change is a constant in the telephone industry, the union and its members must adapt to, rather than block, the introduction of more efficient technologies. Second, the union needed to negotiate retraining programs to teach the current workforce the skills necessary to move into new jobs. Third, the union should negotiate contract provisions such as reduced work hours and pay increases to ensure that the CWA-represented workforce realized the benefits of productivity-enhancing new technologies. In the 1950s, CWA President Joseph Beirne advocated negotiated solutions (an annual guaranteed wage, retraining, shorter work week, longer vacations, better pensions, and higher wages associated with productivity gains) and public policy to respond to automation. By the late 1970s, CWA leaders and key staff, influenced by the Swedish-based socio-technical school, added a fourth element to the CWA technology policy: contractual provisions giving the union a role in the introduction and implementation of new technology. For early response to automation, see Beirne in *The Challenge of Automation,* 67–71; for a critique, see Resnikoff, *Labor's End,* 69–70, 89, 111.

67. Michael D. Dymmel, "Technological Trends and Their Implications for Jobs and Employment in the Bell System," November 19, 1979, Technology folder, CWA research department; CWA's 1980 Bell System Bargaining Council Resolution, box 106, folder 1, CWA-TL.

68. For discussion of the implementation of the Quality of Work Life program, see chapter 4. The CWA-Bell System Settlement: 1980, CWA research department; CWA Letter to Members, Summary of 1980 CWA-Bell Settlement, August 18, 1980, box 106, folder 8, CWA-TL.

69. George Kohl, interview with author. Kohl began his career at CWA in 1980 and became CWA research director a few years later. He served as assistant to CWA Presidents Morton Bahr, Larry Cohen, and Christopher Shelton until his retirement from CWA in 2018. Throughout his career, Kohl had major responsibility for research support for collective bargaining, organizing, and public policy; 1983 CWA Final Bargaining Report and Letter of Understanding, CWA District 2 1983 CWA Final Bargaining Report and Letter of Understanding Notebook, CWA District 2.

Chapter 2. Becoming a Workforce of Resistance

1. Mazzeo, interviews with author.

2. Letter accompanying bargaining proposal submitted by Judy Buchanan and thirteen coworkers, presumably 1986, 1986 C&P Local Bargaining Minutes Cont. Notebook, CWA District 2/13 office.

3. Frederick Winslow Taylor designed his system of scientific management in the first decades of the twentieth century as a method to boost productivity, increase worker wages, and in his view, promote harmonious labor-management relations. Using time-motion studies, managers "scientifically" studied the most efficient method to perform each discrete task in the labor process, appropriating workers' knowledge to break down the production process into discrete tasks, and regulating the manner and speed at which workers did their jobs. For Taylor, the supervisors' job was to do the thinking while the worker did the work. Many companies adopted elements of Taylor's system to deskill, speed up, and wrest control over the labor process from workers. Contrary to Taylor's stated goals, his methods fomented labor-management strife and degradation of labor standards and conditions. The Bell system adopted Taylor's system of time-motion studies to manage the telephone operators. On scientific management, see Taylor, *Principles of Scientific Management*. Many critical assessments of Taylorism exist. A good place to start is Braverman, *Labor and Monopoly Capital*, especially 124–38. On the Bell system adaptation of Taylorism in the operator workplaces, see Green, *Race on the Line*, 73–76, 117–36.

4. Government policy plays a critical role in shaping the economic context in which unions and employers contend for power in the collective bargaining process. The greater the market power of a firm, the greater its profits and resources available to share with its workforce. Conversely, the weaker the firm's market power, the more fiercely management will fight with the union about sharing its resources. Katz, Kochan, and Colvin, *Labor Relations in a Globalizing World*, 79–101, 82–83.

5. Vietor, *Contrived Competition*, 211.

6. Temin and Galambos, *Fall of the Bell System*, 109–12, 217–77 (citation on 277); Schlesinger, Dyer, Clough, and Landau, *Chronicles of Corporate Change*, 141; Cole, *After the Breakup: Assessing the New Post-AT&T Divestiture Era*, 1–18; AT&T Annual Report, 1983.

7. Temin and Galambos, *Fall of the Bell System*, 344 (for quote). For an alternative view that argues technology drove regulatory change in the telecommunications industry, see Vietor, *Contrived Competition*, 167–233, and Derthick and Quirk, *Politics of Deregulation*. On the role of big business in telecommunications reform, see Schiller, *Telematics and Government*, 97–188; Vogel, *Fluctuating Fortunes*, 169.

8. Levinson, *An Extraordinary Time*, 65–81; Kotz, *Rise and Fall of Neoliberal Capitalism*, 14–44, 63–64, 85–126; McCraw, *Prophets of Regulation*, 228, 237; Vietor, *Contrived Competition*, 11–12; Vogel, *Fluctuating Fortunes*, 113–36; Burgin, *The Great Persuasion*.

9. Breyer, *Regulation and Its Reform*; Kahn, *Economics of Regulation*; MacAvoy, *Regulated Industries and the Economy*; Stigler, "The Theory of Economic Regulation"; Derthick and Quirk, *Politics of Deregulation*, 8–224.

10. Airline Deregulation Act of 1978; Motor Carrier Act of 1980 (trucking); Railroad Revitalization and Regulatory Reform Act of 1976, followed by the Staggers Rail Act of 1980 (railroads). Vietor, *Contrived Competition*, 23–90.

11. Vietor, *Contrived Competition*, 15–16; Temin, *Fall of the Bell System*, 53, 264–65, 284–87, 312–16.

12. Author's calculation of Bell system job loss from company SEC Form 10-K, various years; see also Keefe and Boroff, "Telecommunications Labor-Management Relations"; Temin and Galambos, *Fall of the Bell System*, 365.

13. Alt, Collins, Kramer, Leonard, Schaffer, and Mazzeo, interviews with author.

14. In an important article, legal scholar Hiba Hafiz critiques regulators' failure to consider the impact of divestiture remedies on workers' countervailing power, using the labor market impacts of the AT&T breakup as a prime example to make her case. While Hafiz acknowledges that dismantling dominant firms can result in more firms competing for workers' services, which can lift wages, she notes, "It can also dismantle structures of worker power that have arisen to successfully counter dominant employers. A leading example is the devastating effect of the breakup of the Bell System in the 1980s on the Communications Workers of America, gutting union density within the telecommunications industry. . . ." Hafiz convincingly argues that regulators' "failure to include labor market competition protections as a component of their remedial design and, at worst, their imposition of remedial protections *reduce* labor market competition." Hafiz urges contemporary antitrust agencies to analyze the impact of structural remedies on the labor market, in general, and on workers' institutions of collective power. See Hafiz, "Rethinking Breakups," 1491.

15. AT&T and Bell Atlantic SEC Form 10-K, various years; Keefe and Boroff,

"Telecommunications Labor-Management Relations," 303–71; Batt, "Performance and Welfare Effects of Work Restructuring: Evidence from Telecommunications Services," 88–89.

16. For AT&T and Bell Atlantic Employment and Union Data, 1984–2002, see Tables A1 (AT&T) and A2 (Bell Atlantic) in the appendix.

17. I borrow the phrase "irony of telecommunications reform" from Horwitz, *Irony of Regulatory Reform*.

18. Crandall, "Surprises from Telephone Deregulation and the AT&T Divestiture"; Temin and Galambos, *Fall of the Bell System*, 308–17, 345; Vietor, *Contrived Competition*, 14–220. For FCC deregulation of AT&T long-distance service, see *Motion of AT&T Corp. to be Reclassified as a Non-Dominant Carrier Order*, 11 FCC Rcd 3271, October 23, 1995 (rel).

19. AT&T's share of long-distance minutes fell from 80.2 percent in 1984 to 55 percent in 1995; AT&T's share of long-distance carrier revenue dropped from 90 percent to 52 percent over the same period. FCC, *Statistics of Communications Common Carriers*, 1996/7 edition, Table 8.8 (minutes), Table 1.4 and 1.5 (revenues). For equipment market, see Huber, Kellogg, and Thorne, *The Geodesic Network II*, Table 6.2, 6.12; Egan and Waverman, "The State of Competition in Telecommunications."

20. *AT&T 1995 Annual Report*, February 11, 1996, 6; Edmund Andrews, "AT&T Acquisition, Soon to Be Spun Off, Retains NCR Name," *The New York Times*, January 11, 1996, D5.

21. For annual financial performance data, see Table A3 (AT&T) and Table A4 (Bell Atlantic) in the appendix. Source: AT&T and Bell Atlantic SEC Form 10-K, various years.

22. For an overview of the FCC's Computer I (1996), Computer II (1976), and Computer III (1985) proceedings, see Canon, "The Legacy of the Federal Communications Commission's Computer Inquiries." For an early vision of the public benefits of the intelligent network, see *The Intelligent Network Task Force Report*, reprinted as Appendix I in *Pacific Bell's Response to the Intelligent Network Task Force Report*, 1988 (in author's possession).

23. Office of Technology Assessment, *The Electronic Supervisor*, 35.

24. AT&T marketed its PBX equipment and multiple software products to call center operators through a newsletter, the *AT&T Consultant Exchange*. "ISDN gateway products streamline call centers operations," *AT&T Consultant Exchange* 3, no. 6 (December 1989): 10–11; "DEFINITY System G3 provides powerful support for call center applications," *AT&T Consultant Exchange*, February 1992; "AT&T Advanced Routing Solutions, Optimizing Your Call Center Performance," nd; AT&T Call Center Solutions, "AT&T Advanced Operations Portfolio," nd; "Call Centers: The '90s Approach to Customer Service," boxes 44-10-03, 390-03-03, 11-10-03-05, AT&T Archives.

25. Karl, "Panorama of Collective Bargaining," 161; Memo from H.B. Pierson, CWA National Director, to All Long Lines Local Presidents re Service Representative, December 15, 1980, box 114, folder 16, CWA-TL; Communications

Workers of America Recognition List, 1983, CWA research department; Stake, interview with author.

26. Author calculation based on CWA data. Active Domestic Occupational Employees as of 12/09/95 in CWA Telecommunications and Technology "Time in Title" notebook, CWA Telecommunications and Technologies Department.

27. About 25 percent of the employees at the Columbus consumer center was African American. Schmitz, interview with author. About 40 percent of the Pittsburgh billing office was African American. Schaffer, interview with author. In contrast, the Syracuse business center was 95 percent white. Mazzeo, interview with author.

28. AT&T SEC Form 10-K for the year ended December 31, 1983, 3.

29. Rosemary Batt's study in the year 2000 of more than 350 customer service centers in the telecommunications industry found greater autonomy and less surveillance in call centers serving business customers compared to those serving consumer customers. Batt, Colvin, Katz, and Keefe, *Telecommunications 2000*.

30. The seven mega centers were in Pittsburgh, PA, Charleston, WV, Atlanta, GA, Kansas City, MO, Lee's Summit, MO, Dallas, TX, and Mesa, AZ. "Customer Sales and Service: Lee's Summit Center" slide deck, February 22, 1995, CWA research department; Leonard, interview with author; Project Omega notes of George Kohl, CWA research director, nd. For data on the number and locations of the AT&T consumer centers before mega center consolidation, see Letter from C.D. Andrews, AT&T District Manager-Labor Relations, to D.E. Treinen, Assistant to the Vice President, CWA, Attachment A, September 23, 1985; "Customer Centers to Reorganize," *AT&T Focus,* November 22, 1988, 15; AT&T Press Release, "Customer Service Center in Pittsburgh to Add 180 Positions," August 31, 1993. "World-Class Service," *AT&T Focus,* February 1992 (16 centers).

31. The consolidated business centers (known as Incoming Call Receipt Centers, ICRCs) were located in Syracuse, NY, New Orleans, LA, Portland, OR, and Worthington, OH. CWA Commercial/Marketing Conference, April 1996, CWA research department.

32. Schmitz, interview with author. Schmitz was a Force and Facilities manager in the Columbus, OH, CSSC and transferred to the Lee's Summit CSSC when the Columbus center closed in 1992.

33. Mazzeo, interview with author.

34. Letter from Robert H. Livingston, AT&T Director of Labor Relations, to James E. Irvine, CWA Vice President, February 12, 1985. Time in Title Locations in C&T Notebook, CWA Telecommunications and Technologies (T&T) office.

35. Letter from Arthur L. Harris, Adm. Asst. to CWA Vice President Irvine to Chere Chaney, President, CWA Local 6450, May 28, 1996, CWA T&T office.

36. Stipulation #46, Customer Service and Billing Organization Titles—

August 12, 1984, CWA T&T office stipulation notebook; Mazzeo and Unger, interviews with author. Unger began her career at AT&T in 1979 as a communications technician, a beneficiary of the 1973 EEOC consent decree. She rose through the ranks of her New York City Local 1150 to become president in 1987, leading a local of thousands of mostly male technicians. She also represented service order administrators, a customer service title, until they were largely eliminated through automation. Unger was elected to the bargaining team for the AT&T Communications contract beginning in 1992 to represent the "Marketing" employees, and she served on each bargaining team through 2005. She joined CWA staff in 2008 and became assistant to the vice president in CWA's Telecommunications and Technologies division.

37. The arbitrator designated $5 million for wage zone consolidation, allocated $85,000 to the IBEW, leaving $4.9 million for CWA, which he split between the second-tier title and information service assistants. CWA Title/Wage/Zone Presentation, nd (presumably around 12/15/2000 per handwritten notation), CWA T&T files; Decision of Daniel G. Collins, Arbitrator, February 1, 2001, CWA T&T files; CWA, 1986 Bargaining Resolution, ATT-C Bargaining Unit, CWA offsite archives, box 904001; "ATTIS/ATTCOM Convention," March 10–16, 1986 Notebook, 6; Harris Letter to Chere Chaney, May 18, 1996, CWA T&T files; Alt, interview with author.

38. In the two years after divestiture, Bell Atlantic cut 5,423 non-management employees (9.5 percent of the non-management workforce) and 5,337 management employees (21.8 percent of the management workforce). "Bargaining '86—Preparing for Tomorrow," presumably 1986, CWA District 2 1986 bargaining notebook, CWA District 2/13 office in Lanham, MD.

39. "Proposal: Job Title and Pay of Service Representative Be Upgraded to Reflect the Past and Ongoing Changes of Job Responsibilities and the Constant Increase in Duties," nd, CWA District 2 1986 C&P Local Bargaining Minutes Cont. Notebook.

40. "Job Description: RASC Service Representative," submitted by Judy Buchanan, Sandra Deavers, Rita Dooley, Sue Fulton, Brownie Haracivet, Jackie Knight, Marti Lowrie, Julie Martel, Marva Potts, Frances Randall, Sheri Renn, Delores Rowe, Linda Surber, Nettie Womack, presumably 1986, CWA District 2 1986 C&P Local Bargaining Minutes Cont. Notebook.

41. Letter from Frances C. Randall, Service Representative, To Whom It May Concern (and CCed to C&P management and CWA Local 2205 President James Stroup), Newport News, VA, June 11, 1986; Untitled bargaining proposal submitted by 28 signed service representatives, nd, CWA District 2 1986 C&P Local Bargaining Minutes Cont. Notebook.

42. CSR/SR Settlement Agreement, Letter from Hazel Dellavia, president CWA Local 1024 and member of the bargaining committee, to CWA Brothers and Sisters, June 9, 1987, CWA District One Trenton, NJ, office.

43. Letter of Agreement Re: Collector Title from H.A. Clark Jr., New Jersey Bell Director of Labor Relations, to Hazel P. Dellavia, CWA Staff

Representative, July 1992, CWA District One 1992 Local Bargaining Notebook—New Jersey Bell, CWA District 1 Trenton, NJ, office.

44. Consultant Agreement Letter of Understanding, New Jersey Bell Commercial Marketing Letters of Agreement Notebook, CWA District 1 Trenton office, signed February 14, 17, and 28, 1994, by Barbara (Lephardt) Fox Shiller, assistant to vice president CWA District 2; Hazel Dellavia, area representative CWA District 1; JoAnn Diana, president CWA Local 1023; G.P. Dreves, staff representative, CWA District 13; Sandra Kmetyk, executive vice president CWA Local 13500, and signed March 8, 1994, by Ron Williams, director, Bell Atlantic NSG Labor Relations.

45. "Practices and Procedures for Bell's Sale of Optional Services," April 10, 1990, Exhibit F, CWA Local 13500 Office, OCA Action Against Bell of PA folder.

46. Hamm and Alt, interviews with author.

47. Mazzeo, Alt, and Downing, interviews with author. Mary Ann Alt rose from steward and vice president to be elected president of Local 2150 in Maryland (1991–2001). Mary Ellen Mazzeo served as steward, vice president, and then president of CWA Local 1152 in Syracuse, NY, (1997–2013) and represented the commercial/marketing group on the AT&T Communications national bargaining committee in every negotiation from 1998 through 2018. The large call centers resembled factories in the size of the union membership. There were two AT&T mega centers in the Kansas City/Lee's Summit, MO, area. CWA Local 6450 had as many as 10,000 employees in the 1990s, led by women local presidents (Cherie Chaney, Judi Stearns, and Colleen Downing). Other female local presidents with large consumer call centers included Billie Gavin of Local 6150 in Dallas, TX, and Annie Rogers of Local 7050 in Mesa, AZ.

48. CWA Commercial/Marketing conference agendas and notes from plenary and break-out sessions, 1992–2000, CWA research department files. Key leaders of the early Commercial/Marketing conferences included Sandy Kmetyk from Local 13500 in Pennsylvania, Melissa Morin from Local 1400 in New England, Susan Ryke from Local 7777 in Denver, CO, and Kathy Kinchius from Local 9415 in Oakland, CA.

49. For a good summary of the relationship between scientific management and workplace monitoring, see OTA, *Electronic Supervisor*, 17–18. See also Meyer, *The Five-Dollar Day*; Montgomery, *Workers' Control in America*; Porter Benson, *Counter Cultures*.

50. CWA, *1997 CWA Stress Survey of Commercial/Marketing Employees*, August 1997, CWA research department files.

51. The AT&T contractual agreement states: "[N]o employee shall be disciplined as a result of service sampling [another term for supervisory observation or monitoring] except for gross customer abuse, fraud, violation of privacy of communications, or when developmental efforts have not been successful." CWA/AT&T Contract, Other Operations Agreements from Paragraph (c) of

the 1992 Settlement Memorandum, Item SS, 392–3, May 31, 1992; Dowling, interview with author. Dowling began her career with Ohio Bell as an operator in 1979; she transferred to AT&T when the operator center became an AT&T center after divestiture. She became an account representative in the Columbus, OH, CSSC in 1990, which closed in 1992 when she transferred to the Lee's Summit, MO, CSSC where she worked until 2019. She became Local 6450 unit vice president in 1999 and was elected local president in 2005, serving in this office through 2012.

52. Author's recollection of the 1995 Commercial/Marketing conference, which she attended. CWA, *Adhere This, Adhere This: Big Brother Is Watching You* (May 1995), 20, ii. The members of the adherence task force (with their local in parentheses) were Kathy Ciner (L1105), Barbara Mulvey (L2106), Susan Goodson (L3510), Lori Everts (L4900), Alma Diemer (L6507), Carla Floyd (L7901), Joanie Johnson (L9416), Vicky Kintzer (L13500), and the author (CWA research department).

53. CWA, *Adhere This, Adhere This.*

54. CWA, *Adhere This, Adhere This.*

55. CWA, *Adhere This, Adhere This*, 19; Memo from Dina Beaumont, Exec. Assistant to the President, to Bargaining Chairs re: Bargaining on Adherence, May 26, 1995, CWA research department.

56. Adherence flyer for mobilization activities, CWA research department.

57. Adherence flyer and stickers for mobilization activities, CWA research department.

58. American Compensation Association, "The Elements of Sales Compensation." (Stating "Achieving 100 percent of quota should provide total target earnings. About 65 percent of all sales personnel should at least achieve Target Total Compensation.")

59. CWA/AT&T Contract, May 31, 1992, Appendix 4, 277–82. Mazzeo and Unger, interviews with author.

60. Memo from Arthur L. Harris, Adm. Asst. to Vice President to All Local Presidents—Operations Bargaining Unit re Small Business Markets Incentive Compensation Plan, June 28, 1996 and Memo from Arthur L. Harris, Adm. Asst. to Vice President to All Local Presidents—Operations Bargaining Unit re TAAS Oversight Committee Report, June 27, 1997, CWA T&T office files; CWA/AT&T Agreement, Small Business Markets Incentive Compensation Plan, Appendix 4, 1995; Mazzeo, interview with author; Opinion and Award, Arbitration between AT&T and CWA Local 1152, Company Arbitration No. A99-336, Union File No. 99-CC-1152-026, October 9, 2001.

61. Incentive Compensation Proposal, 1995 Labor Negotiations, Response to Questions of June 17, 1995, Presentation, July 18, 1995.

62. US West SEC Form 10-K for the year 1995 filed March 29, 1996. Qwest Communications bought US West in July 2000. CenturyLink bought US West in 2011 and has since renamed the company Lumen.

63. Hill, interview with author. Annie Hill, a beneficiary of the EEOC

Consent Decree, began working for Pacific Northwest Bell in 1972 as a technician. She rose through the CWA ranks to become president of CWA Local 7909, administrative assistant to District 7 vice president and chief bargaining agent with US West, vice president of District 7, and secretary-treasurer of the national union (2011–2015). Despite her work experience as a technician, Hill, like former technician Laura Unger, became a CWA leader on customer service worker issues in her bargaining unit and nationally.

64. In the Matter of Arbitration between Communications Workers of America, union, and Qwest (formerly dba US West Communications), Employer, Re: Reasonable Sales Objectives, CWA Case #7-99-50, Union's Post-Hearing Brief, January 31, 2001, 13 (hereafter Union Brief).

65. Union Brief, 14.

66. Union Brief, 16–17, 19–20.

67. Union Brief, 46–48.

68. Union Brief, 38–46.

69. The classic study of the relationship between job control and stress is Robert A. Karasek Jr., "Job Demands, Job Decision Latitude, and Mental Strain: Implications for Job Redesign." Data and quotes are from 1997 CWA Stress Survey of Commercial/Marketing Employees, August 1997, CWA research department.

Chapter 3. Organizing to Block the Low Road Path

1. Liliette Jaron, Testimony at San Francisco CA Public Forum, U.S. Department of Labor, Bureau of International Labor Affairs, *Official Report of the Proceedings*, February 27, 1996, 71–77.

2. Decision, Gerald A. Wacknov, Administrative Law Judge, Before the National Labor Relations Board, Division of Judges, San Francisco Branch Office, LCF, Inc. d/b/a La Conexion Familiar and Sprint Corporation and Communications Workers of America, AFL-CIO, District Nine, Case 20-CA-26203, August 30, 1995.

3. For the debate among labor scholars about the strengths and weaknesses of the NLRA, see McCartin, "'As Long as There Survives': Contemplating the Wagner Act after Eighty Years," and responses by Cobble, Becker, and Stone in *Labor: Studies in Working Class History*, 21–42, 43–59.

4. Voss and Sherman, "Breaking the Iron Law of Oligarchy"; Milkman and Voss, *Rebuilding Labor*.

5. A large body of literature documents the aggressive corporate assault on unions in the decades since the 1970s. Major works include Weiler, *Governing the Workplace*, 112, 238–239, and n. 18; Freeman Medoff, *What Do Unions Do?*, 1984, and the 2007 update Freeman, "What Do Unions Do: The 2004 M-Brane Stringtwister Edition"; Windham, *Knocking on Labor's Door*; Bronfenbrenner, "Final Report on the Effects of Plant Closing or Threat of Plant Closing on the Right of Workers to Organize."

6. Katz, Batt, and Keefe, "The Revitalization of the CWA"; Bronfenbrenner and Hickey, "Changing to Organize." Other scholars of union revitalization have largely ignored CWA's "bargain to organize" strategy, focusing instead on the Service Employees International Union, Hotel and Restaurant Employees, and United Food and Commercial Workers campaigns led by social movement-oriented organizers to unionize low-wage service workers. See Voss and Sherman, "Breaking the Iron Law of Oligarchy."

7. Temin and Galambos, *Fall of the Bell System*, 131–42; FCC, *Statistics of Common Carriers*, 1996/7 edition, Tables 8.18 and 8.11 (on AT&T market share decline); Keefe and Boroff, "Telecommunications Labor—Management Relations after Divestiture" (on AT&T job loss); Rosemary Batt, "Performance and Welfare Effects of Work Restructuring"; Author's calculation from AT&T SEC Form 10-K, various years (on union density).

8. CWA research department, "Regional Bell Operating Company Employment," nd; Batt, "Performance and Welfare Effects: Evidence from Telecommunications," 88; Keefe and Boroff, "Telecommunications Labor—Management Relations after Divestiture."

9. CWA research department archives, box 599126; CWA membership reports, 1992; Joan H. Kloepfer, *Inside Sprint Corp*.

10. "Getting Organized: Voice of the CWA District 4 Organizing Network," 2:2, April 1991, box 599075, CWA organizing department archives. The Sprint campaign followed a failed 1986 CWA organizing effort at MCI when that company shut down a suburban Detroit call center to block the unionization campaign. The origins of the Jobs with Justice organization began in CWA's efforts to build a labor/community coalition to fight the MCI shutdown. Cohen, "Introduction: Stand Up! Fight Back!" 3–4.

11. "Getting Organized" newsletter; "Honoring the Leadership and Determination of Larry Cohen," CWA Convention Resolution 75A-15-1, Proceedings and Index of the 75th Annual Convention, June 8–10, 2015; Cohen, interview with author.

12. CWA Memo, "National Sprint Organizing Strategy," nd, box 599075, CWA organizing department archives; Cohen, interview with author.

13. "US Sprint Union-Free Management Guide," box 599126, CWA research department archives; Tunja Gardner, Testimony to the Commission on the Future of Worker-Management Relations, September 15, 1993, box 609424, CWA organizing department archives; CWA, "Network: The Newsletter of the Sprint Employee Network," September 1993, box 599126, CWA research department archives.

14. Sprint memo, July 1993, box 599124, CWA research department archives; CWA, "Wired for Justice: A Newsletter for Friends of Sprint Workers," May 1993, box 599126, CWA research department archives; CWA, "We're Sprint Customer Agents and We're Organizing a Union. Here's Why," box 599120, CWA research department archives.

15. CWA, *We're High-Tech and Low-Wage: Labor Costs at Sprint Long-Distance.*

16. "Sprint Organizing—February 1994," box 599124, CWA research department archives.

17. CWA, "Worker Abuse at Sprint: The Case of La Conexion Familiar—One Year after the Closing: Where are the Workers?" nd; Adler, "Breaking La Conexion"; Dora Vogel, Testimony, Official Report of Proceedings before the United States Department of Labor, Bureau of International Affairs Public Forum, San Francisco, February 27, 1996, 50–55.

18. Stipulation of Facts Number 1 in LCF, Inc. d/b/a La Conexion Familiar and Sprint Corp., Case 20-CA-26203, November 8, 1994, box 609248, CWA organizing department archives; CWA, "Worker Abuse at Sprint: Sprint's Anti-Union Campaign at La Conexion Familiar," nd; Decision of Judge Gerald A. Wacknov, August 30, 1995, box 599123, CWA research department archives.

19. Decision of Gerald A. Wacknov, 14–19; Brief of Communications Workers of America District 9 and CWA Local 9410 in Support of Exceptions to the Decision of the Administrative Law Judge, LCF, Inc. d/b/a La Conexion Familiar and Sprint Corporation and Communications Workers of America, AFL-CIO, District 9, Case 20-CA-26203, August 30, 1995; Declaration of Sandra Rusher attached to CWA unfair labor practice charge, CWA v. Sprint/La Conexion Familiar, July 18, 1994, box 599123, CWA research department archives.

20. CWA, "Worker Abuse at Sprint: One Year After the Closing," CWA research department files; Pattee, "Sprint and the Shutdown of La Conexion Familiar."

21. Robert H. Miller, Regional Director, National Labor Relations Board, Petitioner v. LCF, Inc., d/b/a/ La Conexion Familiar and Sprint Corporation, September 22, 1994; Memorandum of Points and Authorities in Support of Petition for Injunction under Section 10(j) of the National Labor Relations Act; National Labor Relations Board Region 20 Press Release, "San Francisco Office of the NLRB to Seek Injunctive Relief to Compel La Conexion Familiar, a subsidiary of Sprint, to reopen its San Francisco Facility," September 16, 1994.

22. CWA, "Worker Abuse at Sprint: Prominent Supporters of Workers' Rights"; CWA advertisement, "A Message to Candice Bergen from Fellow Sprint Employees," August 9, 1994; Letter from President Bill Clinton to Morton Bahr, March 3, 1997; Memo from Rick Braswell, re meeting with Deutsche Telekom/DPG delegation to Sprint, April 23, 1997, box 609248, CWA organizing department archives.

23. Complaint Filed by the Union of Telephone Workers of the Republic of Mexico with the National Administrative Office of the United States of Mexico, reprinted in Complaint Against Sprint Filed by Mexican Telephone Workers Union, *Daily Labor Report*, No. 28, February 10, 1995, D27; Lowe, "The First American Case under the North American Agreement for Labor Cooperation"; CWA Press Release, "In First Mexican Complaint Under NAFTA Against a

U.S. Corporation, Telecom Union Blasts Sprint's Labor Practices," February 9, 1995, box 599124, CWA research department archives.

24. U.S. Department of Labor, "Secretary of Labor Accepts Ministerial Consultations," June 16, 1995; U.S. Department of Labor, "Agreement Reached between Mexico and U.S. on Labor Issue," December 17, 1995; CWA Press Release, "Sprint Shutdown Prompts 3-Nation Investigation into Violations of Workers' Organizing Rights," December 18, 1995, box 599124, CWA research department archives; Official Report of Proceedings before the United States Department of Labor, Bureau of International Affairs, Public Forum, February 27, 1996; Carey Goldberg, "U.S. Labor Making Use of Trade Accord It Fiercely Opposed," *The New York Times*, February 28, 1996, A11.

25. Secretariat of the Commission for Labor Cooperation, "Plant Closings and Labor Rights: The Effects of Sudden Plant Closings on Freedom of Association and the Right to Organize in Canada, Mexico and the United States," June 9, 1997; Kate Bronfenbrenner, "Final Report: The Effects of Plant Closing or Threat of Plant Closing on the Right of Workers to Organize," Submitted to the Labor Secretariat of the North American Commission for Labor Cooperation, September 30, 1996, box 599124, CWA research department archives.

26. Decision of Gerald A. Wacknov, August 30, 1995; CWA Press Release, "Judge Affirms Charges of Illegal Worker Abuse by Sprint Corp. During Union Drive at Sprint/La Conexion Familiar in San Francisco: Union Will Urge Labor Board to Strengthen Remedy and Cite Sprint for Firing of 235," August 31, 1995, box 599124, CWA research department archives.

27. National Labor Relations Board Decision and Order, LCF, Inc. d/b/a La Conexion Familiar and Sprint Corporation and Communications Workers of America, AFL-CIO, District 9, Case 20-CA-26203, December 27, 1996 (322 NLRB 774, 1996 WL 742383); CWA Press Release, "Sprint Corp. Ordered to Rehire and Pay Back Wages to 177 Latino Telemarketers Fired in 1994 for Union Organizing: Labor Board Panel in Washington Backs CWA Appeal over Shutdown of Sprint/La Conexion Familiar in San Francisco," December 30, 1996, box 599124, CWA research department archives; Opinion for the Court, LCF, Inc. d/b/a La Conexion Familiar and Sprint Corporation, Petitioners/Cross-Respondents v. National Labor Relations Board, Respondent/Cross-Petitioner, Communications Workers of America, Intervenor, No. 96–1500, November 25, 1997 (29 F.3d 1276, 327 U.S. App. D.C. 164).

28. In 2023, the UAW used a "bargain to organize" strategy, including a historic Stand Up strike, to win commitments from Ford, GM, and Stellantis to bring thousands of electric vehicle and battery jobs into the national agreements. UAW Press Release, "UAW Members Ratify Historic Contracts at Ford, GM, Stellantis," November 20, 2023; Lerner, interview with author; Sneiderman and Lerner, "Making Hope and History Rhyme"; Katz, *Shifting Gears* (on UAW bargain to organize).

29. Southwestern Bell changed its name to SBC, Inc., in 1994. I use SBC to refer to the parent corporation.

30. Southwestern Bell Corporation, Annual Reports to Shareholders, 1992 and 1993; Southwestern Bell Telephone Company SEC Form 10-K for the year ended December 31, 1993; Communications Workers of America, *CWA at Southwestern Bell: Five Years to Card Check.*

31. CWA, *Five Years to Card Check.*

32. CWA, *Five Years to Card Check,* 13.

33. CWA, *Five Years to Card Check,* 5, 12; CWA Leaflet, "Southwestern Bell Mobile Systems Has Increased Profitability in Every Way," nd; CWA research department analysis, "Union Makes the Difference: A Comparison of Benefits," nd, CWA research department files.

34. CWA, *Five Years to Card Check,* 7.

35. CWA, *Five Years to Card Check,* 7–8.

36. CWA, *Five Years to Card Check,* 10–18.

37. CWA, *Five Years to Card Check,* 13–18.

38. CWA, *Five Years to Card Check,* 5–19.

39. CWA, *Five Years to Card Check,* 21–23.

40. CWA, Proceedings of the 58th Annual Convention, 1997.

41. See chapter 6 and conclusion for discussion about CWA organizing at Verizon Wireless and T-Mobile.

Chapter 4. False Promises? Job Redesign through Union-Management Partnerships

1. Mulligan, interview with author. Linda (Armbruster) Mulligan began her career in the Bell system in 1971 in Florida, later moving to Denver, Colorado, where she became a service representative with Mountain Bell. She remained a service representative for almost 25 years until her retirement in 2002. She rose from steward to executive vice president of CWA Local 7777 in Denver, where she served on the US West bargaining teams in 1995, 1998, and 2000, and as cochair of the CWA/US West job redesign team. From 2001 to 2014, she worked for the AFL-CIO as a field representative.

2. Batt, "Performance and Welfare Effects of Work Restructuring."

3. Victor Reuther quote from foreword to Parker and Slaughter, *Choosing Sides,* v; see entire volume for Parker/Slaughter critique of labor-management participation; see also Wells, *Empty Promises*; Parker, *Inside the Circle*; Parker, "Industrial Relations Myth and Shop-Floor Reality"; for the counterargument, see Bluestone and Bluestone, *Negotiating the Future.*

4. Morton Bahr, statement in AT&T and CWA, *Report on the Workplace of the Future Conference* (presumably 1993), 3; CWA Executive Board, *Executive Board Report on Union-Management Participation for the Telecommunications Industry.* For a discussion of the "union-empowering model of participation,"

see Banks and Metzgar, "Participating in Management." The journal includes an article by CWA President Morton Bahr, "Mobilizing for the '90s," 59–65.

5. Armbruster and Anderson, interviews with author. Anderson began her career as a New Jersey Bell service representative in 1972 and served as steward and then treasurer of CWA Local 1023. She represented CWA on the job redesign committee as part of the Bell Atlantic Mega Team union-management project. In 1999, she was appointed by CWA to serve as union coordinator of the joint CWA/Verizon retiree health benefits team.

6. For a good description of the origins and impact of German co-determination agreements, see McCaughey, "The Codetermination Bargains." For a discussion of the impact of Scandinavian and German works councils and bargaining institutions in telecommunications, see Doellgast, *Exit, Voice, and Solidarity*, and Doellgast, *Disintegrating Democracy at Work*.

7. Extensive literature from this period focuses on workplace restructuring. Key works include Piore and Sabel, *The Second Industrial Divide*; Appelbaum and Batt, *The New American Workplace;* Heckscher, Maccoby, Ramirez, and Tixier, *Agents of Change*; Kochan, Katz, and McKersie, *The Transformation of American Industrial Relations*; Marshall, "Unions, Work Organization, and Economic Performance," 287–315.

8. AFL-CIO, *The New American Workplace: A Labor Perspective*. Major unions that negotiated labor-management partnerships in this period included the United Auto Workers, CWA, the United Steelworkers, and the Amalgamated Clothing and Textile Workers. Union representation increased the success rate of employee participation programs by ensuring that the programs would benefit workers. Levine and Tyson, "Participation, Productivity, and the Firm's Environment."

9. President Bill Clinton articulated his support for the new American workplace in an endorsement blurb for Irving and Barry Bluestone's book, *Negotiating the Future: A Labor Perspective on Business*. "The Bluestones offer a New Covenant for labor and management based on participation, cooperation, and teamwork," he wrote, all of which are necessary "if America is to regain its competitive edge." My appreciation to Nelson Lichtenstein for this citation from his book with Judith Stein, *A Fabulous Failure: The Clinton Presidency and the Transformation of American Capitalism*.

10. Robert Reich in *Report on the Workplace of the Future Conference*, 11–14. For Reich's general argument, see Reich, *The Next American Frontier* and *The Wealth of Nations*.

11. In addition to Chair John Dunlop, members of the commission included Douglas A. Fraser, former UAW president; Paul A. Allaire, Xerox CEO; Kathryn C. Turner, Standard Technology CEO; labor relations scholars and practitioners Thomas A. Kochan, Paula B. Voos, Paul C. Weiler, William B. Gould, and Richard B. Freeman; former secretaries of labor and commerce Juanita Kreps, W.J. Usery, and F. Ray Marshall; and June Robinson representing the Department of Labor.

12. U.S. Department of Labor and U.S. Department of Commerce, U.S. Commission on the Future of Worker-Management Relations, *Fact-Finding Report*, May 1994, 36. http://digitalcommons.ilr.cornell.edu/key_workplace/276; Statement of Douglas A. Fraser, Report of U.S. Commission on the Future of Worker-Management Relations, U.S. Department of Labor and U.S. Department of Commerce, U.S. Commission on the Future of Worker-Management Relations, *Report and Recommendations*, December 1994, 13–14 (in opposition to Dunlop's quid pro quo). For a discussion of the Dunlop Commission and the short-lived nature of labor-management programs, see Nelson Lichtenstein and Judith Stein, *A Fabulous Failure: The Clinton Presidency and the Transformation of American Capitalism*, chapter 9.

13. See chapter 1, 41.

14. CWA/AT&T National Committee on Joint Working Conditions and Service Quality Improvement, "Statement of Principles on Quality of Work Life," presumably 1981, QWL folder, CWA research department. The origins of QWL in the United States date to the 1970s, when the well-publicized auto worker rebellion at the GM Lordstown plant galvanized union leaders and progressive managers to look for alternative models of work design. Irving Bluestone, chief UAW negotiator with General Motors, was an early advocate of quality of work life joint programs. Bluestone and Bluestone, *Negotiating the Future*; see also Heckscher et al., *Agents of Change*, 26. The best study of Lordstown is Moberg, "Rattling the Golden Chains: Conflict and Consciousness of Auto Workers"; see also Cowie, *Stayin' Alive*, 42–49.

15. U.S. Department of Labor, "Quality of Work Life: AT&T and CWA Examine Process after Three Years," 1985.

16. Laura Unger, speech to the United Nations Association of the USA, "Q.W.L. An Anti-Union Strategy," May 12, 1989, in author's possession.

17. Jeffrey Miller, "The Bossless Office: Unique Arizona Experiment Proves Workers Can Run the Show," *CWA News*, February 1984, 6–7; Batt, "Performance and Welfare Effects of Work Restructuring," 98.

18. Testimony of Morton Bahr to Commission on the Future of Worker-Management Relations, September 15, 1993, 2; Kohl, interview with author; Batt, "Performance and Welfare Effects of Work Restructuring," 97.

19. CWA Executive Board Report on Union-Management Participation for the Telecommunications Industry, 1994, CWA research department.

20. CWA/AT&T 1992 Contract, "Workplace of the Future."

21. Morton Bahr, speech to CWA/AT&T Workplace of the Future conference, "A Report on the Workplace of the Future Conference," 3; CWA, "Union Involvement in the Workplace of the Future: A Guide for Staff and Local Unions Representing AT&T Employees," March 1993, 1, Workplace of the Future folder, CWA research department.

22. Heckscher et al., *Agents of Change*, 35–40; Batt, "Performance and Welfare Effects of Work Restructuring," 35–146.

23. AT&T and CWA, "A Report on the Workplace of the Future Conference."

24. Laura Unger, speech for Workplace of the Future forum, February 7, 1995 in author's possession.

25. Unger, interview with author; Unger, speech for Workplace of the Future Forum.

26. Dowling, Schaeffer, Alt, Irvine, Stake, and Leonard, interviews with author; Heckscher et al., *Agents of Change*, 41–45. For a review of the program, see Charles Heckscher, Sue Schurman, Adrienne Eaton, and Beth Craig, "Work Place of the Future: A Research Report," New Brunswick, NJ: Rutgers University Department of Labor Studies and Employment Relations, May 15, 1997, CWA research department files.

27. Stake, interview with author; Customer Service Transformation, "Meeting in a Box" PowerPoint, October 24, 1995, in author's possession; CWA Telecommunications and Technologies Time-in-Title Notebook.

28. The customer representative who serviced small mom-and-pop business customers would now handle nearly all customer needs, including selling, pricing, product/promotion, ordering, billing, inquiry adjustments, and repair referrals. For these low-revenue customers, some transactions would take place using an interactive voice response system (IVR) without any human interaction. Customer service representatives servicing midsize customers who purchased both voice and some data services and typically interacted with AT&T more frequently would provide customers all the services described above, except repair referrals. AT&T assigned a dedicated customer service representative to those customers who interacted with the company on a weekly basis. The customer service representatives who serviced large, national (or even global) business customers who purchased advanced data and virtual network services, typically under terms of a negotiated contract, also provided one-stop shopping to their assigned, dedicated customer. "Meeting in a Box" PowerPoint.

29. Stake, interview with author; author calculation of wage differentials based on CWA/AT&T contract wage schedules.

30. Schaffer and Alt, interviews with author.

31. Morton Bahr, "What's the Long-Term Cost of Short-Term Profits?" *Quality Progress*, 58–58; AT&T SEC Form 10-K for the year ended December 31, 1999; AT&T, "Strategy Overview," AT&T 1998 Annual Report.

32. Mazzeo, interviews with author.

33. Bruce Gordon biography, "The History Makers," July 14, 2013, https://www.thehistorymakers.org/biography/bruce-gordon; Nadirah Sabir, "Keeping the Lines of Communication Open: Bruce Gordon Answers the Call at Bell Atlantic," *Black Enterprise*, May 1, 1995. Bell Atlantic SEC Form 10-K for the year ended December 31, 1993; "The MEGA-TEAM: A New Partnership Focused on the Future," Bell Atlantic Special News Bulletin, presumably 1993, Local 13500 Archives; Kmetyk, interview with author.

34. The steering committee also included Sandy Kmetyk, whom we met in chapter 1, representing the statewide customer service Local 13500 in

Pennsylvania; Jo Ann Diana, dynamic leader representing the customer service locals at New Jersey Bell; and Jane Sutter, the Bell Atlantic vice president for consumer sales. "Consumer Services Mega-120 Teams," Sue Anderson notebook, Tab 1 (for list of teams) and all tabs (for meeting agendas). Sue Anderson, member of the consultant job redesign team, provided the author with a notebook documenting meetings, conference calls, and other materials assembled as part of the job redesign team (hereafter Anderson notebook).

35. Anderson conference notes, Anderson notebook, Tab 2 (Oct 19–20, 1983).

36. Anderson, interview with author; Mega Team 1 Vision, Anderson Notebook, Tab 1; Job Redesign team minutes, August 26, 1993, February 3, 1994, January 27, 1994, Anderson notebook.

37. Mega Team 1 Job Redesign Meeting Minutes, September 22, 1994, Anderson notebook, Tab 7 (cross-trained model), Tab 5, February 2, 1994 (universal design and functional design models). The job redesign team did make progress on seventeen pilot initiatives to improve consultant work processes in the collections, sales, and service centers. Partnership Press Special Edition newsletter, Mega Team 1 Update, nd, Anderson notebook.

38. Memo from Mega 1 Union Members to Mega 1 Team re True Partnership?, May 5, 1994; Mega-30 Meeting minutes, May 19, 1994, Anderson notebook, Tab 7; Kmetyk, Mulvey, Anderson, and Fox Shiller, interviews with author; Anderson notebook, Tab 8 (for discussion of service representative relief package); for Gordon appointment to senior management, see *Wall Street Journal*, May 26, 1994, B10.

39. Bloomberg Business News, "Sweeping Cutbacks Expected at Bell Atlantic," *New York Times*, August 13, 1994; Morton Bahr, "A Message for Ray Smith—And Others," *CWA News*, January 1995.

40. Rather than go on strike, CWA launched an inside/outside mobilization campaign including workplace actions; rallies; outreach to shareholders, elected officials, and the public; and radio, TV, and print advertising from contract expiration in August 1995 to agreement on a contract in February 1996. The contract included substantial wage and pension improvements, no health care givebacks, and provisions to give the union access to jobs in new lines of business (except wireless). "CWA Mounts 'Corporate Campaign' against Telecom Renegade," *CWA News*, September 1995; "Seas of Red Confront Bell Atlantic as CWA Presses Corporate Campaign," *CWA News*, October 1995; "Bell Atlantic Pact Meets the Industry Pattern," *CWA News*, February 1996.

41. Hilton, Kolsrud, Blair, and Buyrn, *Pulling Together for Productivity*; U.S. Department of Commerce and Department of Labor, *Workplace of the Future: A Report of the Conference on the Future of the American Workplace*.

42. Union EIQC members included Reed Roberts (Phoenix), Karla Floyd (Portland, Oregon), Randy Warner (Salt Lake City), and Jim Mahoney (St. Paul). Kevin Boyle from CWA brought seven years of experience in joint union-management processes, and Winnie Nelson, a US West service quality consultant, brought technical expertise in methods of quality improvement.

Hilton et al., *Pulling Together for Productivity.* My analysis draws heavily on the Office of Technology Assessment report and interviews I conducted in 1992–1994, which I updated in drafting this book.

43. *Pulling Together for Productivity*, 42–43.

44. *Pulling Together for Productivity*, 41–45, 53–54. See also Linda Malloy, US West VP-Customer Services, Randy Warner, President CWA Local 7704, Davie Piette, US West Director-Customer Services, "How to Manage without Employee Performance Appraisals," presumably 1993, CWA research department.

45. *Pulling Together for Productivity*, 46–51, 57–61.

46. *Pulling Together for Productivity*, 61; Anthony Ramirez, "US West to Eliminate 9,000 Jobs in 3 Years," *New York Times*, September 18, 1993, 33; "US West Jobs Outlook," *New York Times,* May 4, 1992; Letter from Margaret Hilton, Office of Technology Assessment, to George Kohl, CWA research director, October 28, 1993, CWA research department; Memo from the author to M.E. Nichols, CWA executive vice president, May 20, 1994, CWA research department.

47. Mulligan, interview with author; CWA Press Release, "CWA Strikes US West to Preserve Job Standards, Quality Services, Health Security for Workers and Families," August 16, 1998.

48. Rosemary Batt, "Who Benefits from Teams?"; Batt, "Performance and Welfare Effects of Work Restructuring." Despite the demise of the participation projects described in this chapter, CWA continued to work with management on joint committees. At SBC Communications, a joint service rep task force recommendation was incorporated into a memorandum of understanding that guaranteed no discipline for failure to meet adherence benchmarks. At Ameritech, CWA and call center managers convened a monthly joint committee. At AT&T, joint committees addressed subcontracting (see chapter 5) and provided oversight over the commission plan (see chapter 2).

49. Batt, "The Financial Model of the Firm, the 'Future of Work,' and Employment Relations."

50. Strauss, "Worker Participation—Some Under-Considered Issues."

51. Doellgast, *Disintegrating Democracy at Work*; Doellgast, *Exit, Voice, and Solidarity*.

Chapter 5. Fighting for Job Security

1. Letter from Linda Kramer to Larry Mancino, CWA District 1 vice president, June 8, 1997, CWA research department.

2. Weil, *The Fissured Workplace,* 1–5 (citation on 2); Bernard and Fort, "Measuring the Multinational Economy."

3. Weil, *The Fissured Workplace,* 7–27, 76–121; Arne Kalleberg, *Good Jobs/Bad Jobs*; Doellgast and Panini, "The Impact of Outsourcing on Job Quality

for Call Center Workers in the Telecommunications Industry and Call Centre Industries"; Sweeney and Nussbaum, *Solutions for the New Work Force*, 55–74.

4. Jefferson Cowie, *Capital Moves*; Sugrue, *The Origins of the Urban Crisis*; Gelles, *The Man Who Broke Capitalism*; Bluestone and Harrison, *The Deindustrialization of America*; Winant, *The Next Shift*.

5. Weil, *Fissured Workplace*, 7–27; Kalleberg, *Good Jobs, Bad Jobs*, 27–39; Capelli et al., *Change at Work*, 32–40.

6. Chandler, *The Visible Hand* generally, and 195–206 for discussion of AT&T organizational structure; Chandler and Tedlow, *The Coming of Managerial Capitalism*.

7. Zuboff, *The Age of Surveillance Capitalism*, 40–41.

8. Batt, "The Financial Model of the Firm, the 'Future of Work"; Weil, *The Fissured Workplace*, 43–75; Rodgers, *Age of Fracture*, 41–76; Krippner, *Capitalizing on Crisis*; Lazonick, "Profits without Prosperity"; Davis, *Managed by the Markets*; Kalleberg, *Good Jobs/Bad Jobs*; Appelbaum and Batt, *Private Equity at Work*.

9. Hiatt and Jackson, "Union Survival Strategies for the Twenty-first Century"; Miscimarra and Schwartz, "Frozen in Time—The NLRB, Outsourcing, and Management Rights"; Lerner, Hurst, and Adler, "Fighting and Winning in the Outsourced Economy: Justice for Janitors at the University of Miami." For discussion of the NLRA framework's limitations in the modern fissured workplace, see Andrias, "Constructing a New Labor Law for the Post-New Deal Era"; Madland, *Re-Union*, 1–8. For a comparative study of union organizing in the fissured workplace in the U.S. and Norway, see Franco, "Organizing the Fissured Workplace." For a discussion of global unions' organizing in fissured workplaces, see Drahokoupil, ed., *The Outsourcing Challenge*.

10. For example, since 2012, the "Fight for 15 and a Union" has spearheaded labor/community alliances to raise the minimum wage in states and localities. As of January 2023, twenty-three states and the District of Columbia raised their minimum wage above the federal statutory minimum. Fightfor15 website, https://fightfor15.org/23-states-have-already-raised-their-minimum-wage/, visited June 27, 2023. In 2022, California passed legislation (AB 257) establishing a statewide council to set minimum working standards for fast-food workers. Suhauna Hassain, "A New Law Could Raise Fast-Food Wages to $22 an Hour—And Opponents Are Trying to Halt It," *Los Angeles Times*, September 12, 2022. Public-sector unions have adopted this strategy in their campaigns to block contracting out of government work to non-union, lower-wage private contractors.

11. Letter from Raymond Williams to CWA President Morton Bahr, May 27, 1989. The letter is printed in the 1998 CWA/AT&T Contract, May 10, 1998, 357.

12. Telecommunications Act of 1996, Pub. LA. No. 104–104, 110 Stat. 56 (1996).

13. The Telecommunications Act of 1996 recognized the potential for market failure in serving low-income customers, high-cost rural areas, schools, and libraries. It established universal service mechanisms of subsidies for voice service to low-income households, operating subsidies to rural telephone companies, and subsidies to schools, libraries, and rural health centers. These universal service policies, however, did not provide a mechanism to ensure universal investment in next-generation digital networks, leading to the market failure that is still prevalent today. See U.S. Code Section 254 (universal service). In 2022, 8.3 million U.S. households, largely in rural areas, did not have access to high-speed internet (defined as speeds of 25/3 mps down/up load), and almost one-quarter of American consumers did not subscribe to broadband, largely because of the high cost. FCC, "National Broadband Map: It Keeps Getting Better," May 30, 2023, https://www.fcc.gov/national-broadband-map-it-keeps-getting-better (for access data); Pew Research Center, *Mobile Technology and Home Broadband, 2021* June 3, 2021, 3 (for subscribers). To address these gaps, Congress included $65 billion to support broadband deployment and adoption in the 2021 Infrastructure Investment and Jobs Act. For a summary of those provisions, see Casey Lide, "An Overview of Broadband Provisions in the Infrastructure Bill (as of July 30, 2021)," *The National Law Review*, August 2, 2021, https://www.natlawreview.com/article/overview-broadband-provisions-infrastructure-bill-july-30-2021.

14. The Clinton administration and Representative Markey believed that competition would drive cable and telephone companies' investment in new digital technologies, just as the competitive entry of direct broadcast satellite television spurred cable companies to upgrade their networks in the wake of the 1992 Cable Act. Irving, head of the National Telecommunications and Information Administration in the Clinton administration, and Crowell, chief communications aide to Chairman Markey, interviews with author. See also Edmund Andrews, "Communications Reshaped: The Overhaul; Congress Votes to Reshape Communications Industry, Ending Four Year Struggle," *The New York Times*, February 2, 1996; Crowell, "The Twentieth Anniversary of the Telecommunications Act of 1996."

15. The Telecommunications Act of 1996 envisioned initial competitive entry into local markets through long-distance and cable company resale of service delivered by leasing all or portions of the RBOCs' networks. Therefore, the legislation offered the RBOCs the carrot of entry into long-distance markets only after they demonstrated to federal and state regulators that they had opened their networks to interconnection through extensive investment in new ordering and provisioning systems. In addition, FCC and state regulators set the prices at which the RBOCs were required to lease access to their networks. Initially, these resale rates were set low based on a forward-looking methodology that assumed the most efficient network costs rather than the historic costs the RBOCs had sunk into their networks. 47 U.S.C. Section 271 (special provisions concerning Bell Operating Companies).

16. AT&T and Bell Atlantic SEC Form 10-K, various years.

17. AT&T, 2000 Annual Report, March 19, 2001 ("Three years ago it was clear that technology and regulation would transform the telecommunications industry. AT&T had to act. We needed to move beyond long-distance. So we improved the margins of our core business and used the cash flow to fund our own transformation."). Bell Atlantic SEC Form 10-K for the year ended December 31, 1993, 5 ("Most of the funds for these [broadband network platform] expenditures are generated internally").

18. AT&T SEC Form 10-K for the year ended December 31, 2000; Bell Atlantic SEC Form 10-K for the year ended December 31, 2000.

19. AT&T SEC Form 10-K for the years ended December 31, 2001, and December 31, 2004. The local telephone companies had relationships with almost every household and business in their local territories and, particularly after RBOC consolidation, could provide a significant portion of long-distance service on their own network at minimal additional cost. This was especially true in the Bell Atlantic region that included the large cities in the northeastern and mid-Atlantic United States. Verizon SEC Form 10K for the years ended December 31, 2000, and December 31, 2004.

20. AT&T SEC 2005 Form 10-K for the year ended December 31, 2004. It was only in the mid- to late 2000s that wireless and cable succeeded in challenging the Bell companies' dominance in voice service; cable also emerged as the dominant provider of broadband Internet service.

21. In 1995, CWA supported TransTech employees seeking union representation for the 700 workers employed directly by the company (another 4,300 workers in the TransTech call center were employed by temporary agencies). AT&T waged an anti-union campaign and CWA lost the election. CWA Memo from Morton Bahr, president, and James Irvine, CWA vice president, to All C&T Local Presidents re Special Alert for Stewards Mobilization, February 2, 1995, CWA organizing department archives. Convergys was formed in 1998 from the merger of AT&T's TransTech and two other telemarketing firms, Mattrixx Marketing and Cincinnati Bell Information Services. By 2001, Convergys reported forty-seven call centers and 46,000 employees. Convergys SEC Form 10-K for the years ended December 31, 1998, and December 31, 2001.

22. CWA first negotiated the limitation on contracting in 1989. The letter of agreement that prohibits contracting that causes layoffs or part-timing of work has been included in AT&T and RBOC contracts since that date. Letter from Raymond Williams to CWA President Morton Bahr, May 27, 1989. The letter is printed in the 1998 CWA/AT&T Contract, May 10, 1998, 357.

23. CWA CSSC Proposals, "CWA Proposals to Maintain and Grow Bargaining Unit Work while Maintaining the Competitiveness of AT&T Consumer," December 5, 2001, AT&T Contracting 2000 folder, CWA research department; Stake, interview with author.

24. Union committee members were Colleen Dowling (Local 6450, Lee's Summit, MO), Annie Rogers (Local 7050, Mesa, AZ), Billie Gavin (Local 6150,

Dallas, TX), Clarence King (Local 6143, San Antonio, TX), Martin Quintanilla (Local 6733, El Paso, TX), Gary Allen (Local 1051, Fairhaven, MA), and Lois Grimes and Jerry Klimm, CWA staff representatives. Company members included Steve Leonard, Joe Scuderi, and Joan Gallagher. The author and financial consultant Randy Barber provided research support to the union. Barber was an early advocate of union leveraging of pension funds as a source of power, see Rifkin and Barber, *The North Will Rise Again*. See discussion of early proponents of labor union pension activism in Jacoby, *Labor in the Age of Finance*.

25. The head count figures in the AT&T internal consumer centers are from "Headcount: Internal v External 1999–2001 spreadsheet," AT&T Contracting 2000 folder, CWA research department. In addition to the CSSC call center employees, AT&T consumer also included operators, telemarketers, and other union-represented employees. The eight consumer call centers handling calls regarding domestic service were located in Lee's Summit, MO, Dallas, TX, Mesa, AZ, Sacramento, CA, Kansas City, KS, Fairhaven, MA, Charleston, WV, and Pittsburgh, PA; the three call centers handling calls for international service were in New York City, San Antonio, TX, and El Paso, TX.

26. All the vendor call centers were located in the United States. The operators were Convergys (seven centers), Aegis (three centers), Precision Response Corporation, and TCIM. CWA analysis, "Headcount: Internal v External, 1999–2001," September 27, 2001; CWA analysis, "Call Volume, by Type and Channel, Jan. 1999–Jan. 2004, March 22, 2004," AT&T Contracting 2000 folder, CWA research department.

27. CWA Analysis, "Impact of On-Line/Off-Line and Cost/Quality Adjustments on Unit Cost Differential between Internal and External Sites," November 8, 2001, AT&T Contracting 2000 folder, CWA research department.

28. Four factors explain the higher service quality in the in-house centers: in-house employees had higher rates of first call resolution (service representatives resolved the customers' problem on the first call), higher rates of bridge to sale (turning a service call into a sale), higher rates of recovery of unpaid bills, and lower rates of inappropriate transfers.

29. The company booked the time that in-house representatives spent on follow-up paperwork as an expense in the numerator of the "cost per call" calculation, thereby boosting the "cost per call" ratio for in-house centers. Outsourced centers did little, if any, follow-up paperwork. Also, the in-house centers did not always operate at full capacity. Barber calculated that increasing the call volume to fill capacity in the internal centers (making more efficient use of sunk resources) would further reduce the unit cost, bringing the gap between internal and external to 8.5 percent to 10 percent. CWA, "Union Proposals to Maintain and Grow Bargaining Unit Work While Maintaining the Competitiveness of AT&T Consumer," December 5, 2001, AT&T Contracting 2000 folder, CWA research department; CWA Analysis, "Impact of

On-Line/Off-Line and Cost/Quality Adjustment on Unit Cost Differential between Internal and External Sites," November 8, 2001, AT&T Contracting 2000 folder, CWA research department.

30. A full 34 million of AT&T's 45 million consumer long-distance customers (76 percent) spent less than $25 on their monthly bill. "AT&T Long Distance Customers and Revenues Bands Spreadsheet," November 8, 2001, AT&T Contracting 2000 folder, CWA research department; "Union Proposals."

31. Letter from William A. Stake, Vice President—Sales & Customer Care, AT&T Consumer, to Ralph Maly, CWA Vice President, December 13, 2001, AT&T Subcontracting 2000 files, CWA research department.

32. The job redesign subcommittee trialed a cross-functional team in the Dallas and El Paso, TX, call centers in 2003. Account representatives spent a portion of every day on the phone and a portion offline doing follow-up paperwork. There were small performance improvements, but AT&T did not continue the pilot. The wellness committee trialed a four-day work week that led to some reduction in absenteeism. The sales incentive committee piloted an incentive pay initiative that paid account representatives a bonus for every sale along with the base wage. See various subcommittee reports, AT&T Contracting 2000 folder, CWA research department.

33. "May—Current Offered spreadsheet," February 3, 2005, AT&T Subcontracting 2000 folder, CWA research department.

34. I discuss CWA strategies to build solidarity with call center workers in the Philippines, Dominican Republic, and other offshore locations in the epilogue. There is significant literature on the global call center industry. A good discussion of the challenges global call center outsourcing poses for U.S. and German telecommunications workers and their unions is Doellgast, *Disintegrating Democracy at Work*, 122–79; see also Batt, Holman, and Holtgrewe, "The Globalization of Service Work"; Padios, "Listening Between the Lines"; Freeman, *High Tech and High Heels in the Global Economy* (for a discussion of women workers in data processing centers in Barbados). In both the Philippines and Barbados, the call center and data processors appreciate their jobs' higher pay and air-conditioned working conditions, but they find the pace of work and management highly stressful.

35. From summer 2001 to January 2005, long-distance monthly call volumes dropped from 4.1 million to just 721,000, a decline of 3.4 million calls per month. AT&T monthly local service calls fell from 1.3 million in June 2001 to 721,000 over this same period. Local service calls took an even steeper nosedive in early 2005 after the Federal Communications Commission (FCC) issued a ruling that raised the wholesale rate for the resale platform, effectively killing the local resale market. AT&T 2000 Contracting folder, CWA research department. FCC, *Unbundled Access to Network Elements Order on Remand*, WC Docket No. 04–313, December 15, 2004 (adopted).

36. "AT&T, Current Offered Spreadsheet" (for February 2005 data) and

CWA spreadsheet, "Call Volume, by Type and Channel, Jan. 1999–Jan 2004," March 22, 2004 (for June 2001 data); CWA Headcount spreadsheet, nd, AT&T Subcontracting 2000, CWA research department. Between 1999 and 2004, AT&T consumer revenue nosedived from $21.7 billion to $7.9 billion. AT&T SEC Form 10K for the years ended December 31, 2000 and December 31, 2005.

37. During this brief period, workers at about twenty AT&T Broadband locations (the cable division) joined CWA under a negotiated expedited election procedure. AT&T delayed negotiations for first contracts in most of these units. After Comcast bought the AT&T cable business in 2002, Comcast management adopted an aggressive anti-union stance and eventually decertified the union in all locations. See American Rights at Work, *No Bargain: Comcast and the Future of Workers' Rights in Telecommunications* (Washington, DC: American Rights, 2004); CWA, "CWA Case Study in Union Busting Show Weaknesses in Labor Law," November 1, 2003, https://cwa-union.org/news/entry/case_study_in_union_busting_comcast_abuses_show_weakness_in_labor_law.

38. T-shirt in author's possession.

39. Bell Atlantic News Release, "New Bell Atlantic Company Brings 'Megacenter,' 700 New Jobs to Virginia; One-Stop Shopping for Anything from Cellular to Internet Access," March 27, 1997, reproduced in CWA Manual, "Bargaining for Our Future: 1998 Bell Atlantic Mobilization," Spring 1998, 3–6, 3–7; Memo from Victoria Kinzer and Sandra Kmetyk to Kris Raab, CWA research economist, October 31, 1997, CWA research department.

40. Author's personal recollection as an observer at the rally; Summerlyn, interview with author.

41. CWA, "Bargaining for Our Future: 1998 Bell Atlantic Mobilization," CWA research department.

42. CWA, "Bargaining for Our Future: 1998 Bell Atlantic Mobilization."

43. Author's notes on conversation with Melissa Morin about Bell Atlantic Plus, June 6, 1998, CWA research department.

44. Letter from Morton Bahr, CWA President, to Donald J. Sacco, Bell Atlantic Executive Vice President for Human Resources, July 15, 1998, Dina Beaumont files, CWA research department.

45. CWA Press Release, "CWA Ends Strike at Bell Atlantic with Settlement Extending Union Recognition over Jobs on the Information Highway," August 11, 1998; CWA Bell-Atlantic South Regional Report #30, August 11, 1998; CWA/Bell Atlantic Memorandum of Agreement, August 11, 1998, 1998 Common Issues Bell Atlantic Memorandum of Agreement Notebook, CWA District 2 Office.

46. Morton Bahr, email communication to author, December 4, 2016.

Chapter 6. Striking for Stress Relief

1. Irwin, interview with author.

2. Simon Romero, "Labor Accord Hits New Economy," *The New York Times*,

August 22, 2000, C1; CWA PowerPoint, "Verizon 2000 Bargaining: A Giant Victory for Labor," presumably fall 2000, box 9124104, CWA research department archives (hereafter CWA Verizon 2000 PowerPoint); 2000 Common Issues Memorandum of Understanding between Verizon and Communications Workers of America, Commercial Stress Relief Package, August 13, 2000, Bell Atlantic 2000 bargaining folder, CWA research department.

3. Labor historians have largely ignored this CWA strike as well as other major CWA strikes against telephone employers (New York Telephone in 1971, AT&T in 1983 and 1986, NYNEX in 1989, Southern New England Telephone in 1998, US West and Bell Atlantic in 1998). See Brenner, Day, and Ness, eds., *The Encyclopedia of Strikes in American History*; Jeremy Brecher, *Strike! Revised and Expanded.* Gabrielle Semel's forthcoming book *The Cablevision War: How a Majority Black Workforce and a Union Willing to Fight Won against a Billionaire* makes an important contribution to filling this gap, with a detailed analysis of militant strike culture in CWA District 1 (which includes New York and New England).

4. The seven bargaining units included the CWA contracts with Verizon North (formerly Bell Atlantic North in New York and New England), Verizon Mid-Atlantic (formerly Bell Atlantic South in New Jersey, Pennsylvania, Delaware, Maryland, Virginia, West Virginia, and Washington, DC), two small CWA construction units, a CWA unit of about 100 workers at Verizon Wireless in New York City, an IBEW unit in New Jersey (covering technicians), and an IBEW unit in New England (covering technicians, operators, and a small group of service representatives). CWA Verizon 2000 PowerPoint.

5. Brecher, Review of *The Encyclopedia of Strikes in American History*. The statistics come from the U.S. Bureau of Labor Statistics database on major work stoppages. The data have some analytical problems. From 1947 to 1981, the BLS recorded strikes of all sizes. After 1981, it recorded only "major work stoppages" involving more than 1,000 workers for at least 24 hours. A careful analysis by L.J. Perry and Patrick J. Wilson concluded that trends in small strikes tended to correspond to those in larger strikes and that one can assume that the overall pattern of decline in the number of strikes is valid. See L.J. Perry and Patrick J. Wilson, "Trends in Work Stoppages: A Global Perspective."

6. Bureau of Labor Statistics, "Work Stoppages Involving 1,000 or More Workers in 2023," (through October 2023) and "Annual Work Stoppages Involving 1,000 or More Workers, 1947–Present"; McCartin, "Solvents of Solidarity"; Brecher, "American Exceptionalism and the 'Death of the Strike.'" For an excellent study of the PATCO strike, see McCartin, Collision Course. For summaries of major strike victories in 2023, see UAW, "UAW Workers Ratify Historic Contracts at Ford, GM, and Stellantis," November 20, 2023, https://uaw.org/uaw-members-ratify-historic-contracts-ford-gm-stellantis/; Writers Guild of America West, "Summary of the 2023 WGA MBA," https://www.wga.org/contracts/contracts/mba/summary-of-the-2023-wga-mba; SAG-AFTRA, "SAG-AFTRA National Board

Approves Tentative Agreement, Recommends Ratification of 2023 TV/Theatrical Contracts," November 10, 2023, https://www.sagaftra.org/sag-aftra-national-board-approves-tentative-agreement-recommends-ratification-2023-tvtheatrical.

7. For Verizon corporate ranking, *see* Fortune 500 online archive, https://money.cnn.com/magazines/fortune/fortune500_archive/full/2001/index.html.

8. Verizon SEC Form 10-K for the year ended December 31, 2020.

9. Bahr, interview with author.

10. For a discussion of the 1971 strike, see Brenner, "Rank-and-File Struggles at the Telephone Company"; Master, interview and correspondence with author; Early and Wilson, "How a Telephone Workers' Strike Thirty Years Ago Aided the Fight for Single Payer"; Brown, "Why We Wear Red on Thursdays"; for an in-depth study of CWA strike activity, see Gabrielle Semel's forthcoming book *The Cablevision War: How a Majority Black Workforce and a Union Willing to Fight Won against a Billionaire.*

11. CWA, *Bargaining for Our Future: 2000 Bell Atlantic Mobilization Manual,* Washington, DC, 2000 Bell Atlantic bargaining folder, CWA research department.

12. CWA Verizon 2000 PowerPoint; CWA Strike Bulletin Issues #3, #6, and #7, August 10, 15, and 16, 2000, Mobilization notebook, 2000 Verizon East bargaining, box 9124104, CWA research department archives.

13. CWA Verizon 2000 PowerPoint; CWA Radio Spot on Stress and Forced Overtime, presumably between August 6–23, 2000, Bell Atlantic 2000 bargaining folder, CWA research department; Mary Williams Walsh, "When 'May I Help You' Is a Labor Issue," *The New York Times,* August 12, 2000, C1; "Verizon and Unions Fail to Agree on Contract," *The New York Times,* August 6, 2000, 22; Deborah Solomon and Yochi J. Dreazen, "Verizon Hit by Strikes, But Talks Progress," *The Wall Street Journal,* August 7, 2000, A3.

14. CWA, Bell Atlantic/Verizon Strike Bulletin Issues 5, 7, and 8, August 13, 16, and 18, 2000. Mobilization notebook, 2000 Verizon East bargaining, box 9124104, CWA research department archives.

15. Barbara Wago, steward Local 13500, email to CWA President Morton Bahr, August 12, 2000. CWA Bell Atlantic/Verizon Strike Bulletin, Settlement Issue, presumably late August 2000, Mobilization notebook, 2000 Verizon East bargaining, box 9124104, CWA research department archives.

16. Wendy Tanaka, "Area Phone Workers Seek a Better Deal, Verizon Employees in Pa., N.J. and States South Stay on Strike, Citing Overtime and Other Issues," *The Philadelphia Inquirer,* August 22, 2000, A01.

17. Cohen, interview with author; Multiple drafts of proposed CWA/Verizon Memorandum of Agreement Regarding Neutrality and Card Check at Verizon from Aug. 6, 2000 to Aug. 18, 2000, Bargaining Notes Notebook, 2000 Verizon-East Bargaining, box 696640, CWA research department archives; CWA/Bell Atlantic 1998 Memorandum of Agreement, August 1998; SBC Memorandum of

Agreement Regarding Neutrality and Card Check Recognition, March 1997, in CWA, *Five Years to Card Check*; "Strikers Win Groundbreaking Agreement," *CWA News,* Sept. 2000, 3. For a discussion of how Verizon Wireless violated the spirit of neutrality, see epilogue.

18. Common Issues Memorandum of Understanding between CWA and Verizon Delaware, Verizon Maryland, Verizon New Jersey, Verizon Pennsylvania, Verizon Services Corp., Verizon Virginia, Verizon Washington, DC, and Verizon West Virginia, August 23, 2000. Bell Atlantic 2000 bargaining folder, CWA research department; CWA District One Final Regional Bargaining Report, Verizon Communications, August 22, 2000; CWA Bell Atlantic/ Verizon Strike Bulletin, Settlement Issue, presumably late August 2000. Mobilization notebook, 2000 Verizon East bargaining, box 9124104, CWA research department archives.

19. "Service Rep Issues for Bell Atlantic Bargaining," memo summarizing author's interviews with CWA local leaders representing customer service workers, presumably January 2020, Verizon 2000 bargaining, Service Rep issues folder, CWA research department. Five of the seven leaders of CWA call center workers that were interviewed were women. The local leaders were Sandy Kmetyk, Local 13500 in Pennsylvania, Linda Kramer, Local 1023 in New Jersey, Barbara Mulvey, Local 2106 in Salisbury, MD, Melissa Morin, Local 1400 in New England, Kenny Rucker, Local 2222 in northern Virginia, and Keith Edwards, Local 1105 in New York metro area.

20. "Service Rep Issues for Bell Atlantic Bargaining." The union sometimes referred to the negotiations and bargaining units that year as Bell Atlantic, other times as Verizon units.

21. Bell Atlantic Data Requests Notebook, Tab 4, data as of April 26, 2000; Bell Atlantic response to CWA June 5, 2000, information request dated June 20, 2000, "June 2000 Early Commercial Marketing Talks Notebook," CWA District 1 Trenton office; Service Rep Issues for Bell Atlantic Bargaining (for the year 2000); Bell Atlantic Associate Count by NCS, Data as of 3/28/1992. Bell Atlantic 1992 Notebook Response to Data Request, box 635892, CWA research department archives; CWA Report to Shareholders at Bell Atlantic, distributed at Bell Atlantic shareholders meeting, May 2000, 2000 Verizon East bargaining, box 9124104, CWA research department archives.

22. CWA Radio Spot on Stress and Forced Overtime, presumably sometime between August 6 and August 24, 2000. 2000 Verizon-East Bargaining, Box 9124104, CWA research department archives; Service Rep Issues for Bell Atlantic Bargaining, Bell Atlantic 2000 bargaining, Service Rep Issues folder, CWA research department.

23. Service Rep Issues for Bell Atlantic Bargaining.

24. CWA, 2000 Bargaining Briefing Paper, "Justice on the Job: Working Conditions in the Customer Service Occupations," presumably March 2000; 2000 CWA Bargaining Council Resolution—Bell Atlantic: Setting the Pace for the New Millennium, presumably March 15, 2000. Both documents in

Dina Beaumont papers, 2000 Bell Atlantic Service Rep Bargaining, tri-district meeting folder; "CWA Mobilizing for Summer Talks with Bell Atlantic," *CWA News*, April 2000.

25. Participants in these meetings included CWA representatives Keith Edwards (Local 1105, NY), Paula Lopez (Local 1105, NY), Melissa Morin (Local 1400, New England), Sandy Kmetyk (Local 13500, PA), Pat Scoville (Local 2202, Virginia Beach, VA), Barbara Mulvey (Local 2106, Salisbury, MD), Linda Kramer (Local 1023, NJ), Patti Chronic (Local 1022, NJ), DJ Bryant (Local 13001, DE), Jim Byrne (District 13 staff), Carol Summerlyn (District 2 staff, former service representative), Pat Niven (District 1 staff), Ed Baxter (District 1 staff). IBEW representatives included Joan Haigh, Margie Scholle, and Mary Jo Arcuri. Company representatives included labor relations staff Jackie Latheram, Bill Drucker, Debbie Dartanell, Ruthie Burton, and Sandy Bousman.

26. Union Proposal for the Company to Relieve Stress, presumably June 1, 2000, Early Commercial Marketing Talks Notebook, CWA District 1 Trenton, NJ, office.

27. Kmetyk, interview with author; Memo from author to Dina Beaumont re Service Rep Issues Committee, presumably late June 2000, Bell Atlantic 2000 folder, CWA research department.

28. Business Office Issues Meeting Notes, CWA District 1 Trenton N.J. office, June 2000 Commercial Marketing Talks notebook, June 5, 2000–July 16, 2000; Memo from author to Dina Beaumont re Service Rep Issues Committee; Memo from Linda Kramer and Patricia Chronic to Lawrence Mancino, CWA District 1 Vice President, re Commercial/Marketing Sub-Committee, July 18, 2000, Bell Atlantic Service Rep Bargaining folder, Dina Beaumont papers.

29. CWA, Commercial Marketing Issues, August 2, 2000; Stress Presentation, August 1, 2000.

30. Kintzer, Summerlyn, and Irwin, interviews with author.

31. Deana Smith and Kim Rogers, emails to and reply from CWA President Morton Bahr, August 11, 2000. Bell Atlantic 2000 Bargaining folder, CWA research department.

32. CWA Radio Spot on Stress and Forced Overtime; CWA Press Release, "Too Much Stress Means Workers Suffer—And So Does Customer Service," August 30, 2000, Bell Atlantic 2000 bargaining folder, CWA research department.

33. CWA Press Release, "Too Much Stress Means Workers Suffer"; CWA flyer, "Don't Let Verizon Hang Up on Good Jobs," 2000 Verizon-East Bargaining, Mobilization Notebook, box 9124104, CWA research department archives; Mary Williams Walsh, "When 'May I Help You' Is a Labor Issue'" *New York Times*, Aug. 6, 2000; Diana Nelson Jones, "Are Service Reps Facing Last Call for Joy on the Job?," *Pittsburgh Post-Gazette*, August 24, 2000; Sarah Schafer, "Home Life on Hold? Verizon Strikers Say Call Centers Take Toll," *Washington Post*, August 18, 2000; Jim McKay, "Phone Strike Causing Hang-Ups," *Pittsburgh Post-Gazette*, August 8, 2000.

34. Memo from Linda Kramer, President, CWA Local 1023, to Dina Beaumont, Executive Assistant to the President, August 16, 2000, Bell Atlantic 2000 bargaining folder, CWA research department.

35. Summerlyn and Kinzer, interviews with author; Verizon Mid-Atlantic bargaining minutes, CWA District 2/13 office in Lanham, MD..

Epilogue

1. Cohen and Sabol, interviews with author. Ed Sabol was organizing director in CWA District 1 and then national organizing director with oversight over the Verizon Wireless campaign until his retirement in 2012. See "Verizon Wireless Anti-Union Animus," compiled May 13, 2008; Verizon Wireless Including Verizon-RMT box, CWA President's Office Archives (at CWA headquarters).

2. Verizon Communications SEC Form 10-K for the year ended December 31, 2022.

3. CWA press releases, "Verizon Wireless Retail Workers in Oswego, Illinois Form Union with Communications Workers of America," November 18, 2022; "Verizon Retail Workers in Portland Vote to Form a Union as Seattle Area Retail Workers Ratify First Contract," September 12, 2022. For NLRB charges against Verizon Wireless, see David Groves, "CWA files multiple charges against Verizon for retaliation," *The Stand,* August 24, 2022; CWA press release, "Verizon Wireless Reinstates Illegally Fired Worker, Provides Compensation for Lost Wages," March 16, 2023; Michael Sainato, "It's Union Busting 101: Documents Reveal Verizon's Attacks on Organized Labor," *The Guardian,* January 16, 2019; Tim Dubnau, CWA District 1 Organizing Coordinator, email to author, June 30, 2021.

4. CWA membership database as of May 2023, Dan Reynolds, CWA assistant research director, email to author (for CWA members at Verizon call centers); Verizon SEC Form 10-K for the year ended December 31, 2022 (for union employment). In 2021, Verizon's wireline service had just over 9 million customers.

5. AT&T SEC Form 10-K for the year ended December 31, 2022.

6. CWA, *AT&T 2018 Jobs Report*; CWA and National Employment Law Project, *Broken Network: Workers Expose Harms of Wireless Telecom Carriers' Outsourcing to 'Authorized Retailers.'"* Wage rates from CWA/AT&T wireline contracts in District 3 (Southeast) and District 9 (California/Nevada) and CWA/AT&T Mobility "orange" contract. Employment figure from CWA membership database as of January 2023.

7. CWA Press Release, "AT&T Mobility Worker Discusses Impact of AI with White House Staff," June 1, 2023; Emma Goldberg, "'Training My Replacement': Inside a Call Center Worker's Battle with A.I.," *The New York Times,* July 19, 2023.

8. Cable workers at Cablevision (now called Optimum) in Brooklyn, New

York, waged a five-year battle to win a union and a contract. CWA Press Release, "Brooklyn Cablevision Workers Get Contract after Three-Year Battle," February 19, 2015; Semel, interview with author. Semel describes the Cablevision organizing and contract campaign in her forthcoming book, tentatively titled *The Cablevision War: How a Majority Black Workforce and a Union Willing to Fight Won against a Billionaire.*

9. The median telecommunications workers' hourly wage was $26.51 in the 1983–1986 period and $27.59 in the 2016–2019 period in constant (inflation-adjusted) dollars. John Schmitt and Jori Kandra, *Decades of Slow Wage Growth for Telecommunication Workers,* Table 5; Jori Kanda, email correspondence with author, June 22, 2021. Author's calculation of productivity change, 1987–2019 from Bureau of Labor Statistics, U.S. Labor Productivity database for NAICS code 5173 (wired and wireless telecom). For union density, see Barry T. Hirsch and David A. Macpherson, "Union Membership, Earnings, and Coverage from the CPS," http://unionstats.com (for 1983); U.S. Bureau of Labor Statistics, Union Members—2022, Table 3, January 19, 2023 (for 2022). The leading broadband providers (Comcast, Spectrum, and Cox) are non-union. Two of the big three wireless companies (Verizon and T-Mobile) are non-union. For T-Mobile violations of labor law, see CWA, "Stop the Systematic Abuse at T-Mobile," 2015; T-Mobile Workers United website, http://tmobileworkersunited.org/about-us/; CWA, "Judge Orders T-Mobile to Disband Illegal Workplace Organization," April 4, 2017. For Comcast, see American Rights at Work, *No Bargain: Comcast and the Future of Workers' Rights in Telecommunications*; CWA, "CWA Case Study in Union Busting Show Weaknesses in Labor Law," November 1, 2003, https://cwa-union.org/news/entry/case_study_in_union_busting_comcast_abuses_show_weakness_in_labor_law.

10. See CWA press releases, "CWA Investigates AT&T Offshoring Operation in the Dominican Republic," May 4, 2017; "CWA Members Protest AT&T Contractor Alorica," April 25, 2019; "CWA and Dominican Union FEDOTRAZONAS Announce Solidarity Agreement to Build Worker Power," December 5, 2019; "UNI ICTS Gives First-Ever Excellence in Organizing Awards to SITT, Sitratel-Fedotrazonas, and GPEU," August 27, 2019; "American, Filipino AT&T Call Center Workers Meet in Manila, Demonstrate Global Solidarity Between Unions," August 26, 2019; "Following Tragic Fire in the Philippines, CWA Calls on AT&T, T-Mobile, and Verizon to Ensure Safety of Workers in Offshored Call Centers," January 25, 2018; "CWA Stands with International Call Center Workers," September 27, 2018; Kalena Thomhave, "American Call Center Workers Rally for Their Filipino Counterparts," *The American Prospect,* May 23, 2019; Marcia Brown, "Filippino Workers Who Refused to Break a Strike Now Jailed for Union Activism," *The American Prospect,* November 21, 2019.

11. Danielle Abril, "Your Boss Can Monitor Your Activities with Special Software," *Washington Post,* October 17, 2022, A21; Norm Scheiber and Niraj Chokshi, "A Rebellion By Workers as Railroads Get Leaner," *The New York Times,* September 16, 2022; Jules Roscoe, "Amazon Workers Are Still Peeing in

Bottles," *Motherboard*, November 2, 2022; Colin Lecher, "How Amazon Automatically Tracks and Fires Warehouse Workers for 'Productivity,'" *The Verge*, April 25, 2019; Jodi Kantor, Karen Weise, and Grace Ashford, "The Amazon That Customers Don't See," *The New York Times*, June 15, 2021; SoftActivity website, https://www.softactivity.com/get/employee-monitoring/?utm_source=bing&utm_medium=ad&utm_campaign=employee%20monitoring, viewed January 9, 2023.

12. CWA press releases, "Proletariat Workers Become Third Group of Activision Blizzard Workers to Form Union with Communications Workers of America," December 27, 2022; "Quality Assurance Workers at Microsoft's ZeniMax Studios Establish Company's First Union with Communications Workers of America," January 3, 2023; "Google Workers, Demanding Change at Work, are Launching a Union with Communications Workers of America," January 4, 2021; "Workers at Penn Square Apple Store in Oklahoma City Celebrate Groundbreaking Win," October 14, 2022; "Workers at TCGPlayer Celebrate Groundbreaking Win, TCGUnion/CWA Becomes the First Certified Union at EBay," March 10, 2023; "Tower Climbers Release Report Exposing Safety Hazards and Mobilize for Bill of Rights to Improve Conditions," May 25, 2023; Katy Robertson, "New York Times Tech Workers Vote to Certify Union, *The New York Times*, March 3, 2022.

13. CWA press releases, "CWA-Microsoft Announce Labor Neutrality Agreement," June 13, 2022; "Qualtek Climbers Vote to Join CWA," May 24, 2022; "IUE-CWA Members Win Historic Agreement for More Union Jobs at GE," May 18, 2023; CWA Organizing Update, April 27, 2023.

14. NLRA, 29 U.S.C. §§ 151.

15. On the PRO Act (H.R. 842), see govtrack, https://www.govtrack.us/congress/bills/117/hr842. On labor law reform proposals, see Block and Sachs, *Clean Slate for Worker Power: Building a Just Economy and Democracy*, 10; Andrias, "Constructing a New Labor Law for the Post-New Deal Era"; David Madland, *Re-Union*.

16. The wealthiest fifty Americans are worth almost $2 trillion, while the total net worth of the bottom half of the U.S. population is just $2.08 trillion. Ben Steverman and Alezandre Tanzi, "The Richest 50 Americans Are Worth as Much as the Poorest 165 Million Americans," *Bloomberg Wealth*, October 8, 2020, citing U.S. Federal Reserve data.

Bibliography

Primary Sources

MANUSCRIPT COLLECTIONS

AT&T Archives and History Center. Warren, New Jersey.
CWA District 2/13 Office. Lanham, Maryland.
CWA Local 13500 Office. Pittsburgh, Pennsylvania.
CWA District 1 Office. Trenton, New Jersey.
CWA Organizing Department, Washington, DC, and Springfield, Virginia (offsite archive).
CWA Research Department. Washington, DC, and Springfield, Virginia (offsite archive).
CWA Telecommunications and Technologies Department, Washington, DC, and Springfield, Virginia (offsite archive).
CWA Presidents' Office. Washington, DC.
National Archives at College Park, Maryland. FCC Records, Docketed Case Files 1927-90, Docket No. 19143, Record Group 173.
Tamiment Library/Robert F. Wagner Labor Archives, NYU Special Collections, New York University. Communications Workers of America Records.

AUTHOR'S INTERVIEWS

Mary Ann Alt. March 9, 2013, March 13, 2013, September 3, 2000.
Clara Sue Anderson. January 25, 2012, December 31, 2020.
Larry Cohen. December 28, 2015.
Ron Collins. December 4, 2011.
Colin Crowell. August 4, 2022.
Beverly Davis. November 1, 2021.
Hazel Dellavia. January 24, 2012, February 10, 2012.

Colleen Downing. May 29, 2020, April 22, 2021.
Gail Evans. January 29, 2012.
Danny Fetonte. December 4, 2016.
Lois J. Grimes-Patow. February 20, 2014.
Michelle Guckert. July 23, 2014.
Glenn Hamm. February 24, 2013.
Elizabeth Hargrove. December 3, 2021.
Yvette Herrera. January 26, 2022.
Annie Hill. October 26, 2020.
James Irvine. January 27, 2013.
Larry Irving. August 8, 2022.
Marilyn Irwin. December 1, 2021.
Leslie Jackson. February 7, 2022.
Victoria Kintzer. May 20, 2021.
Sandy Kmetyk. February 1, 2012, January 21, 2013.
George Kohl. January 12, 2021.
Linda Kramer. January 8, 2012.
Mike T. Kzirian. October 31, 2014.
Steven J. Leonard. March 9, 2014.
Stephen Lerner. December 13, 2022.
Robert Master. May 19, 2021.
Mary Ellen Mazzeo. April 19, 2012, June 5 and 9, 2020.
Linda Mulligan. November 17, 2020.
Sandy Rusher. January 2, 2017.
Ed Sabol. July 29, 2022.
Ameenah Salaam. September 7 and 8, 2022.
Mary Lou Schaffer. May 5, 2013, January 31, 2021.
Jan Schmitz. June 8, 2020.
Gay Semel. December 29, 2022.
Barbara Fox Shiller. January 2, 2013.
William A. Stake. August 21, 2014.
Carol Summerlyn. May 12, 2021.
Angie Thompson, September 2, 2022.
Donald Treinen. April 13, 2020 (email communication).
Laura Unger. August 4, 2020.

CORPORATE DOCUMENTS

AT&T. *Bell System Statistical Manual, 1950–1979,* May 1980.
AT&T Annual Reports and SEC Form 10-K, 1980–2004.
Bell Atlantic Annual Reports and SEC Form 10-K, 1984–2005.
SBC SEC Form 10-K, 1992–1997.
US West SEC Form 10-K, 1996–1999.

CWA COLLECTIVE BARGAINING AGREEMENTS, MINUTES, RESOLUTIONS, AND SUMMARIES

CWA/AT&T Collective Bargaining Agreements. Minutes, Resolutions, and Summaries. 1947, 1960, 1963, 1967, 1971, 1974, 1977, 1980, 1986, 1989, 1992, 1995, 1998.

CWA/Bell Atlantic Common Issues Agreements. Bargaining Reports, Minutes, Resolutions, and Summaries. 1986, 1989, 1995, 2000.

CWA/C&P Collective Bargaining Agreements. 1974, 1980, 1986, 1989, 1992, 1995, 2000, 2003, 2008.

CWA/New Jersey Bell Collective Bargaining Agreements. 1992, 1995, 1998, 2000.

CWA/Pennsylvania Bell Collective Bargaining Agreements. 1968, 1974, 1980, 1983, 1986, 1989, 1992, 1998, 2000, 2003.

CWA/SBC Collective Bargaining Agreement. 1997.

CWA/US West Collective Bargaining Agreements. 1992, 1996, 1998.

NEWSPAPERS

Black Enterprise
Bloomberg Business News
Bloomberg Wealth
CWA News
Motherboard
Pittsburgh Post-Gazette
The Los Angeles Times
The New York Times
The Philadelphia Inquirer
The Verge
The Wall Street Journal
The Washington Post

GOVERNMENT DOCUMENTS

Federal Communications Commission, Washington, DC. *Statistics of Communications Common Carriers; Federal Communications Record,* various broadband reports.

U.S. Bureau of Labor Statistics. *Current Employment Survey.* Employment and Earnings Tables, Union Membership Tables.

U.S. Census Bureau. Current Population Survey, Annual Social and Economic Supplements.

U.S. Senate Hearings before the Subcommittee of Labor Management Relations of the Committee on Labor and Public Welfare on Labor-Management Relations in the Bell Telephone System. Washington, DC: Government Printing Office, 1950.

Secondary Sources

Adler, Bill. "Breaking La Conexion: Sprint Long Distance Pulls the Plug on Its Latino Employees." *The Texas Observer,* September 2, 1994.

AFL-CIO. *The New American Workplace: A Labor Perspective.* Washington, DC: 1994.

American Compensation Association. *The Elements of Sales Compensation.* C5 Self-Study Certification. Edition 5.30. April 18, 1997.

American Rights at Work. *No Bargain: Comcast and the Future of Workers' Rights in Telecommunications.* Washington, DC: American Rights, 2004.

Andrias, Kate. "Constructing a New Labor Law for the Post-New Deal Era." In *The New Deal and Its Legacies,* edited by Romain Huret, Nelson Lichtenstein, and Jean-Christian Vinel, 240–256. Philadelphia: University of Pennsylvania Press, 2020.

Appelbaum, Binyamin. *The Economists Hour: False Prophets, Free Markets, and the Fracture of Society.* New York: Little, Brown, 2019.

Appelbaum, Eileen, and Rosemary Batt. *The New American Workplace: Transforming Work Systems in the United States.* Ithaca, NY: ILR Press, 1994.

Appelbaum, Eileen, and Rosemary Batt. *Private Equity at Work: When Wall Street Manages Main Street.* New York: Russell Sage Foundation, 2014.

AT&T. "DEFINITY System G3 Provides Powerful Support for Call Center Applications." *AT&T Consultant Exchange,* February 1992.

AT&T. "ISDN Gateway Products Streamline Call Center Operations." *AT&T Consultant Exchange* 3, no. 6 (December 1989): 10–11.

AT&T and Communications Workers of America. "A Report on the Workplace of the Future Conference." nd, presumably 1993.

Bahr, Morton. *From the Telegraph to the Internet.* Washington, DC: National Press Books, 1998.

Bahr, Morton. "Mobilizing for the '90s." *Labor Research Review* 18, no 2 (Fall 1989): 59–65.

Bahr, Morton. "What's the Long-Term Cost of Short-Term Profits?" *Quality Progress* (July 1996): 58.

Bain, Peter, and Phillip Taylor. "Entrapped by the 'Electronic Panopticon?' Worker Resistance in the Call Center." *New Technology, Work and Employment* 15, no. 1 (2000): 1–18.

Balanoff, Elizabeth. Oral History Interview with Catherine Conroy. Ann Arbor: University of Michigan, 1978. https://rs4.reuther.wayne.edu/OralHistories/Projects/LOH002227/LOH002227_OH_006.pdf.

Baldry, Chris, Peter Bain, and Phillip Taylor. "Bright Satanic Offices: Intensification, Control, and Team Taylorism." In *Workplaces of the Future,* edited by Paul Thompson, 163–183. London: Macmillan Business, 1998.

Banks, Andy, and Jack Metzgar. "Participating in Management: Union Organizing on a New Terrain." *Labor Research Review,* 18, no. 2 (fall 1989): 1–41.

Barbash, Jack. *Unions and Telephones*. New York: Harper and Brothers, 1952.

Baron, Ava. "Contested Terrain Revisited: Technology and Gender Definition of Work in the Printing Industry." In *Women, Work, and Technology,* edited by Barbara Drygulski Wright. Ann Arbor: University of Michigan Press, 1987.

Baron, Ava. "Gender and Labor History: Learning from the Past, Looking to the Future." In *Work Engendered: Toward a New History of American Labor,* edited by Ava Baron. Ithaca, NY: Cornell University Press, 1991.

Barry, Kathleen M. *Femininity in Flight: A History of Flight Attendants*. Durham, NC: Duke University Press, 2007.

Batt, Rosemary. "Explaining Wage Inequality in Telecommunications Services: Customer Segmentation, Human Resource Practices, and Union Decline." *Industrial and Labor Relations Review* 54, no. 2A (2001): 425–49.

Batt, Rosemary. "The Financial Model of the Firm, the 'Future of Work,' and Employment Relations." In *The Routledge Companion to Employment Relations,* edited by Adrian Wilkinson, Tony Dundon, Jimmy Donaghey, and Alex Colvin. New York: Routledge, 2012.

Batt, Rosemary. "Performance and Welfare Effects of Work Restructuring: Evidence from Telecommunications." PhD diss., M.I.T., 1995.

Batt, Rosemary. "Strategic Segmentation in Front-Line Services: Matching Customers, Employees, and Human Resource Systems." *International Journal of Human Resource Management* 11, no. 3 (2000): 540–61.

Batt, Rosemary. "Who Benefits from Teams? Comparing Workers, Supervisors, and Managers." *Industrial Relations* 43 (January 2004): 183–212.

Batt, Rosemary, Alex Colvin, Harry Katz, and Jeffrey Keefe. "Telecommunications 2000: Strategy, HR Practices, and Performance." Ithaca, NY: Cornell, presumably 2000.

Batt, Rosemary, and Owen Darbishire. "Institutional Determinants of Deregulation and Restructuring in Telecommunications: Britain, Germany, and the United States Compared." *International Contributions to Labour Studies* 7 (1997): 59–79.

Batt, Rosemary, Virginia Doellgast, and Hyunji Kwon. *The U.S. Call Center Industry 2004: National Benchmarking Report*. New York: Alfred P. Sloan Foundation, 2004.

Batt, Rosemary, David Holman, and Ursula Holtgrewe. "The Globalization of Service Work: Comparative International Perspectives on Call Centers." *Industrial and Labor Relations Review* 62, no. 4 (2009): 453–88.

Batt, Rosemary, and Hiroatsu Nohara. "How Institutions and Business Strategies Affect Wages: A Cross-National Study of Call Centers." *Industrial and Labor Relations Review* 62, no. 4 (2009): 533–52.

Batt, Rosemary, Hiroatsu Nohara, and Hyunji Kwon. "Employer Strategies and Wages in New Service Activities: A Comparison of Coordinated and Liberal Market Economies." *British Journal of Industrial Relations* 48, no. 2 (2010): 400–35.

Baumol, William J., Otto Eckstein, and Alfred E. Kahn. "Competition and Monopoly in Telecommunications Services." Reprint in *Hearings before the Subcommittee on Antitrust and Monopoly*. Committee on the Judiciary, 93rd Congress, 1st Session, November 23, 1970.

Beirne, Joseph. *The Challenge of Automation*. Washington, DC: Public Affairs Press, 1955.

Beirne, Joseph. Testimony, Hearings before the Subcommittee on Labor-Management Relations of the Committee on Labor and Public Welfare, United States Senate, Eighty-first Congress, Second Session, August 10, 1950.

Bell, Daniel. *The Coming of Post-Industrial Society: A Venture in Social Forecasting*. New York: Basic Books, 1973.

Belzer, Michael H. *Sweatshops on Wheels: Winners and Losers in Trucking Deregulation*. New York: Oxford University Press, 2000.

Bennett, James T., and Bruce E. Kaufman, eds. *The Future of Private Sector Unionism in the United States*. Armonk, NY: M.E. Sharpe, 2002.

Bennett, James T., and Bruce E. Kaufman, eds. *What Do Unions Do? A Twenty-Year Perspective*. New Brunswick, NJ: Transaction Publishers, 2007.

Benson, Susan Porter. *Counter Cultures: Saleswomen, Managers, and Customers in American Department Stores, 1890–1940*. Urbana: University of Illinois Press, 1986.

Benze, Dorothee. "Organizing to Survive, Bargaining to Organize." *Labor and Society* 5, no. 1 (June 2002): 95–107.

Bernard, Andrew B., and Teresa C. Fort, "Measuring the Multinational Economy: Factoryless Goods Producing Firms," *American Economic Review* 105, no. 5 (2015): 518–23.

Blaxall, Martha, and Barbara Reagan, eds. *Women and the Workplace: The Implications of Occupational Segregation*. Chicago: University of Chicago Press, 1975.

Block, Sharon, and Benjamin Sachs. *Clean Slate for Worker Power: Building a Just Economy and Democracy*. Cambridge, MA: Harvard Law School Labor and Worklife Program, 2019.

Bluestone, Barry, and Irving Bluestone. *Negotiating the Future: A Labor Perspective on American Business*. New York: Basic Books, 1992.

Bluestone, Barry, and Bennett Harrison. *The Deindustrialization of America*. New York: Basic Books, 1982.

Bluestone, Barry, and Bennett Harrison. *The Great U-Turn: Corporate Restructuring and the Polarizing of America*. New York: Basic Books, 1988.

Bonnett, Thomas W. *Telewars in the States: Telecommunications in a New Era of Competition*. Washington, DC: Council of Governors Policy Advisors, 1996.

Boris, Eileen, and Jennifer Klein. *Caring for America: Home Health Workers in the Shadow of the Welfare State*. New York: Oxford University Press, 2012.

Boyle, Kevin, and Sue Pisha. "Building the Future through Quality Unions in the Telecommunications Industry—Communications Workers of America and US West." In *Union, Management, and Quality: Opportunities for Innovation and Excellence,* edited by Edward Cohen-Rosenthal. Boston: Irwin Professional Pub., 1995.

Braverman, Harry. *Labor and Monopoly Capital: The Degradation of Work in the Twentieth Century.* New York: Monthly Review Press, 1974.

Brecher, Jeremy. "American Exceptionalism and the 'Death of the Strike.'" *New Labour Forum* 12, no. 3 (Fall 2003): 98–102.

Brecher, Jeremy. Review of *The Encyclopedia of Strikes in American History* by Aaron Brenner, Benjamin Day, and Immanuel Ness, eds., *Labor: Studies in Working-Class History* 8, no.1 (Spring 2011): 131–32.

Brecher, Jeremy. *Strike! Revised and Expanded.* Oakland, CA: PM Press, 2004.

Brenner, Aaron, Benjamin Day, and Immanuel Ness, eds. *The Encyclopedia of Strikes in American History.* Armonk, NY: M.E. Sharpe, 2009.

Brenner, Aaron, Benjamin Day, and Immanuel Ness. "Rank-and-File Struggles at the Telephone Company." In *Rebel Rank and File: Labor Militancy and Revolt from Below during the Long 1970s,* edited by Aaron Brenner, Robert Brenner, and Cal Winslow. New York: Verso, 2010.

Breyer, Stephen. *Regulation and Its Reform.* Cambridge, MA: Harvard University Press, 1982.

Brock, Gerald. *The Telecommunications Industry.* Cambridge, MA: Harvard University Press, 1981.

Brody, David. *In Labor's Cause: Main Themes on the History of the American Worker.* New York: Oxford University Press, 1993.

Brody, David. *Labor Embattled.* Urbana: University of Illinois Press, 2005.

Brody, David. *Workers in Industrial America: Essays on the 20th Century Struggle.* New York: Oxford University Press, 1980.

Brooks, John. *The Communications Workers of America.* New York: Mason/Charter, 1977.

Brooks, John. *Telephone: The First Hundred Years.* New York: Harper and Row, 1975.

Bronfenbrenner, Kate. "Final Report on the Effects of Plant Closing or the Threat of Plant Closing on the Right of Workers to Organize." Submitted to the Labor Secretariat of the North American Commission on Labor Cooperation. September 30, 1996.

Bronfenbrenner, Kate, and Robert Hickey. "Changing to Organize: A National Assessment of Union Strategies. In *Rebuilding Labor: Organizing and Organizers in the New Labor Movement,* edited by Kim Voss and Rachel Sherman. Ithaca, NY: ILR Press, 2004.

Brown, Marcia. "Why We Wear Red on Thursdays." *The American Prospect,* Oct. 25, 2019.

Burgin, Angus. *The Great Persuasion: Reinventing Free Markets since the Depression.* Cambridge, MA: Harvard University Press, 2012.

Burns, Joe. *Reviving the Strike: How Working People Can Regain Power and Transform America.* Brooklyn, NY: IG Publishing, 2011.

Burroway, Michael. *Manufacturing Consent: Changes in the Labor Process Under Monopoly Capitalism*. Chicago: University of Chicago Press, 1979.
Callaghan, George, and Paul Thompson. "Edwards Revisited: Technical Control and Call Centres." *Economic and Industrial Democracy* 22 (2001): 13–37.
Canon, Robert. "The Legacy of the Federal Communications Commission's Computer Inquiries." *Federal Communications Law Journal* 55, no. 2 (2003): 167–206.
Cappelli, Peter, Lauri Bassi, Harry Katz, David Knoke, Paul Osterman, and Michael Useem, *Change at Work*. New York: Oxford University Press, 1997.
Cassedy, Ellen. *Working 9 to 5: A Women's Movement, a Labor Union, and a Movie*. Chicago: Chicago Review Press, 2022.
Catcher-Gershenfeld, Joel. "Tracing a Transformation in Industrial Relations." U.S Department of Labor-Management Relations and Cooperative Programs, Report No. 123. Washington, DC: Government Printing Office, 1988.
Chandler Jr., Alfred D. *Strategy and Structure in the History of the American Enterprise*. Cambridge, MA: MIT Press, 1962.
Chandler Jr., Alfred D. *The Visible Hand: The Managerial Revolution in American Business*. Cambridge, MA: Harvard University Press, 1977.
Chandler Jr., Alfred D., and Richard S. Tedlow. *The Coming of Managerial Capitalism: A Casebook on the History of American Institutions*. Homewood, IL: Richard D. Irwin, 1985.
Clawson, Dan. *Bureaucracy and the Labor Process: The Transformation of U.S. Industry, 1860–1920*. New York: Monthly Review Press, 1980.
Clawson, Dan, and Mary Ann Clawson, "IT is Watching: Surveillance and Worker Resistance," *New Labor Forum* 26, no. 2 (2017) 62–69.
Coase, Ronald. "The Nature of the Firm," *Economica* 4 (1937): 386–405.
Cobble, Dorothy Sue. *Dishing It Out: Waitresses and Their Union in the Twentieth Century*. Urbana: University of Illinois Press, 1991.
Cobble, Dorothy Sue. *The Other Women's Movement: Workplace Justice and Social Rights in Modern America*. Princeton, NJ: Princeton University Press, 2004.
Cobble, Dorothy Sue, ed. *The Sex of Class: Women Transforming American Labor*. Ithaca, NY: ILR Press, 2007.
Cobble, Dorothy Sue. "'A Spontaneous Loss of Enthusiasm': Workplace Feminism and the Transformation of Women's Service Jobs in the 1970s." *International Labor and Working Class History* 56 (Fall 1999): 23–44.
Cobble, Dorothy Sue. "A Tiger by the Toenail: The 1970s and the Origins of the New Working-Class Majority." *Labor: Studies in Working Class History of the Americas* 2, no. 3 (Fall 2005): 103–14.
Cohen, Jeffrey E. *The Politics of Telecommunications Regulation: The States and the Divestiture of AT&T*. Armonk, NY: M.E. Sharpe, 1992.
Cohen, Larry. "Introduction: Stand Up, Fight Back!" In *Jobs with Justice: 25 Years, 25 Voices,* edited by Eric Larson. Oakland, CA: PM Press, 2013.

Cohen, Lizabeth. *A Consumers' Republic: The Politics of Mass Consumption in Postwar America*. New York: Vintage Books, 2004.
Cole, Barry, ed. *After the Break-Up: Assessing the New Post-Divestiture Era*. New York: Columbia University Press, 1991.
Coll, Steve. *The Deal of the Century: The Breakup of AT&T*. New York: Atheneum, 1986.
Communications Workers of America. *1997 Stress Survey of Commercial/Marketing Employees*. Washington, DC, 1997.
Communications Workers of America. *Adhere This, Adhere This: Big Brother is Watching You*. Washington, DC, 1995.
Communications Workers of America. *AT&T 2018 Jobs Report*. Washington, DC, April 2018.
Communications Workers of America. "Bargaining for Our Future: 1998 Bell Atlantic Mobilization." Washington, DC, 1998.
Communications Workers of America. "CWA Case Study in Union Busting Show Weaknesses in Labor Law." Washington, DC, November 1, 2003.
Communications Workers of America. *Executive Board Report on Union-Management Participation for the Telecommunications Industry*. Washington, DC, 1994.
Communications Workers of America. *CWA at Southwestern Bell: Five Years to Card Check*. Washington, DC, 2015
Communications Workers of America. *CWA History: A Brief Review*. Washington, DC, 1997.
Communications Workers of America. *Stop the Systematic Abuse at T-Mobile*. Washington, DC, 2015.
Communications Workers of America. *We're High Tech and Low Wage: Labor Costs at Sprint Long-Distance*. Report prepared for the U.S. House of Representatives Committee on Government Operations, Washington. DC, March 3, 1993.
Communications Workers of America and National Employment Law Project. *Broken Network: Workers Expose Harms of Wireless Telecom Carriers' Outsourcing to 'Authorized Retailers.'* Washington, DC, February 2023.
Contact Babel. *US Contact Centers: State of the Industry, 2023- 2027*.
Coon, Horace. *American Tel & Tel: The Story of a Great Monopoly*. New York: Longmans Green, 1939.
Copus, David, Lawrence Gartner, Randall Speck, William Wallace, Marjanette Feagan, and Katherine Mazzaferri. "A Unique Competition: A Study of Equal Employment
Opportunity in the Bell System." Washington, DC, 1971.
Cornfield, Daniel B., ed. *Workers, Managers, and Technological Change: Emerging Patterns of Labor Relations*. New York: Plenum Press, 1987.
Cowie, Jefferson. *Capital Moves: RCA's Seventy Year Quest for Cheap Labor*. Ithaca, NY: Cornell University Press, 1999.

Cowie, Jefferson. *Stayin' Alive: The 1970s and the Last Days of the Working Class.* New York: The New Press, 2010.
Crandall, Robert W. *After the Breakup: U.S. Telecommunications in a More Competitive Era.* Washington, DC: Brookings Institution, 1991.
Crandall, Robert W. "Surprises from Telephone Deregulation and AT&T Divestiture." AEA Papers and Proceedings 78:2 (May 1988).
Crompton, Rosemary, and Gareth Jones. *White Collar Proletariat: Deskilling and Gender in Clerical Work.* Philadelphia: Temple University Press, 1984.
Cromwell, Joseph H. *The C&P Story: Service in Action.* Chesapeake and Potomac Telephone Company of Maryland, 1981.
Crowell, Colin. "The Twentieth Anniversary of the Telecommunications Act of 1996," February 7, 2021. https://colincrowell.medium.com/reflections-on-the-25th-anniversary-of-the-telecommunications-act-of-1996-eb588a0fb03c.
Danielian, N.R. *AT&T: The Story of Industrial Conquest.* New York: Vanguard Press, 1939.
Davies, Margery. *Woman's Place is at the Typewriter.* Philadelphia: Temple University Press, 1982.
Davis, Gerald. *Managed by the Markets: How Finance Reshaped America.* New York: Oxford University Press, 2009.
Demaria, Alfred T. *How Management Wins Union Organizing Campaigns.* New York: Executive Enterprises Publications, 1980.
Derthick, Martha, and Paul J. Quirk. *The Politics of Deregulation.* Washington, DC: Brookings Institution, 1985.
Deslippe, Dennis A. *Rights, Not Roses: Unions and the Rise of Working-Class Feminism, 1945–80.* Urbana: University of Illinois Press, 2000.
Diebold Group. *Automation: Impact and Implications. Focus on Developments in the Communications Industry.* Washington, DC, 1965.
Doellgast, Virginia. *Disintegrating Democracy at Work: Labor Unions and the Future of Good Jobs in the Service Economy.* Ithaca, NY: Cornell University Press, 2012.
Doellgast, Virginia. *Exit, Voice, and Solidarity: Contesting Precarity in US and European Telecommunications Industries.* New York: Oxford University Press, 2022.
Doellgast, Virginia. "Negotiating Flexibility: The Politics of Call Center Restructuring in the U.S. and Germany." Ph.D. diss. Ithaca, NY: Cornell University, 2006.
Doellgast, Virginia, and Sean O'Brady. "Collective Voice and Worker Well-Being: Union Influence on Performance Monitoring and Emotional Exhaustion in Call Centers." *Industrial Relations* 60, no. 3 (July 2021), 307–37.
Doellgast, Virginia, and Elissa Panini. "The Impact of Outsourcing on Job Quality for Call Centre Workers in the Telecommunications Industry and Call Center Industries." In *The Outsourcing Challenge: Organizing Workers across Fragmented Production Networks,* edited by Jan Drahokoupil. Brussels: European Trade Union Institute, 2015.

Drahokoupil, Jan, ed. *The Outsourcing Challenge: Organizing Workers across Fragmented Production Networks.* Brussels: European Trade Union Institute, 2015.

duRivage, Virginia. "CWA's Organizing Strategies: Transforming Contract Work into Union Jobs." In *Non-Standard Work: The Nature and Challenges of Changing Employment Arrangements,* edited by Francoise Carre, Marianne A. Ferber, Lonnie Golden, and Stephen A. Herzenberg. Champaign, IL: Industrial Relations Research Association, 2000.

Dymmel, Michael D. "Technological Trends and Their Implications for Jobs and Employment in the Bell System." Washington, DC: Communications Workers of America, November 19, 1979.

Early, Steve. *Save Our Unions: Dispatches from a Movement in Distress.* New York: Monthly Review Press, 2013.

Early, Steve, and Rand Wilson. "How a Telephone Workers' Strike Thirty Years Ago Aided the Fight for Single Payer," *Jacobin,* July 13, 2019.

Edsall, Thomas Byrne. *The New Politics of Inequality.* New York: Norton, 1994.

Edwards, Richard. *Contested Terrain: The Transformation of the Workplace in the Twentieth Century.* New York: Basic Books, 1979.

Elias, Allison. *The Rise of Corporate Feminism: Women in the American Office, 1960–1990.* New York: Columbia University Press, 2022.

Epstein, Gerald A., ed. *Financialization and the World Economy.* Northampton, MA: Edward Elgar, 2005.

Erickson, Ethel. *The Woman Telephone Worker.* Washington, DC: U.S. Department of Labor Women's Bureau Bulletin 286, 1963.

Fagen, M.D., ed. *A History of Engineering and Science in the Bell System: The Early Years 1875–1925.* Murray Hills, NJ: Bell Telephone Laboratories, 1975.

Fantasia, Rick, and Kim Voss. *Hard Work: Remaking the American Labor Movement.* Berkeley: University of California Press, 2004.

Federal Communications Commission. *Unbundled Access to Network Elements Order on Remand.* WC Docket No. 04-313, December 15, 2004.

Fernie, Susan, and David Metcalf. *(Not) Hanging on the Telephone: Payment Systems in the New Sweatshops.* London: Centre for Economic Performance, London School of Economics, 1998.

Fink, Leon, and Brian Greenberg. *Upheaval in the Quiet Zone: A History of Hospital Workers' Union, Local 1199.* Urbana: University of Illinois Press, 1989.

Flanagan, Robert. "Has Management Strangled U.S. Unions?" In *What Do Unions Do? A Twenty-Year Perspective,* edited by James T. Bennett and Bruce E. Kaufman. New Brunswick, NJ: Transaction Publishers, 2007.

Franco, Lucas Albert. "Organizing the Fissured Workplace: The Fight to Cultivate Collective Worker Power in an Era of Nonstandard Work." Ph.D. diss., University of Minnesota, November 2019.

Fraser, Jill Andresky. *White Collar Sweatshop: The Deterioration of Work and Its Rewards in Corporate America.* New York: Norton, 2001.

Fraser, Steve, and Gary Gerstle, eds. *The Rise and Fall of the New Deal Order.* Princeton, NJ: Princeton University Press, 1989.

Freeman, Carla. *High Tech and High Heels in the Global Economy: Women, Work, and Pink-Collar Identities in the Caribbean.* Durham, NC: Duke University Press, 2000.

Freeman, Richard B. *America Works: The Exceptional U.S. Labor Market.* New York: Russell Sage Foundation, 2007.

Freeman, Richard B. "What Do Unions Do: The 2004 M-Brane Stringtwister Edition." In *What Do Unions Do? A Twenty-Year Perspective,* edited by James T. Bennett and Bruce E. Kaufman. New Brunswick, NJ: Transaction Publishers, 2007.

Freeman, Richard B., and James L. Medoff. *What Do Unions Do?* New York: Basic Books, 1994.

Frenkel, Steve, Mark Korczynski, Karen Shire, and May Tam, eds. *On the Front-Line: Organization of Work in the Information Economy.* Ithaca, NY: Cornell University Press, 1999.

Friedman, Milton. *Capitalism and Freedom.* Chicago: University of Chicago Press, 1962.

Friedman, Milton. "The Social Responsibility of Business Is to Increase Its Profits." *The New York Times Magazine,* September 13, 1970, Section SM, 17.

Friedman, Sheldon, Richard W. Hurd, Rudolph A. Oswald, and Ronald L. Seeber, eds. *Restoring the Promise of American Labor Law.* Ithaca, NY: ILR Press, 1994.

Gabin, Nancy. *Feminism in the Labor Movement: Women and the United Auto Workers, 1935–1975.* Ithaca, NY: Cornell University Press, 1990.

Garnet, Robert. *The Telephone Enterprise: The Evolution of the Bell System's Horizontal Enterprise.* Baltimore, MD: Johns Hopkins University Press, 1985.

Garrett, William A. *Phonemanship: The Newest Concept of Marketing.* New York: Farrar, Straus, and Cudahy, 1959.

Garson, Barbara. *Electronic Sweatshop: How Computers Are Transforming the Office of the Future into the Factory of the Past.* New York: Simon and Schuster, 1988.

Geismer, Lily. *Left Behind: The Democrats' Failed Attempt to Solve Inequality.* New York: Public Affairs, 2022.

Gelles, David. *The Man Who Broke Capitalism: How Jack Welch Gutted the Heartland and Crushed the Soul of Corporate America—and How to Undo His Legacy.* New York: Simon and Schuster, 2022.

Gerstle, Gary. *The Rise and Fall of the Neoliberal Order: America and the World in the Free Market Era.* New York: Oxford University Press, 2022.

Gerstle, Gary, Nelson Lichtenstein, and Alice O'Connor, eds. *Beyond the New Deal Order: U.S. Politics from the Great Depression to the Great Recession.* Philadelphia: University of Pennsylvania Press, 2019.

Goldman, Debbie. "Curbing Big Brother in the Workplace: The Campaign of CWA Women against Monitoring of Telephone Operators, 1967–1980," paper submitted in partial fulfillment of M.A. degree, University of Maryland, 2005 (available at Robert E. Wagner Archives, Tamiment Library, New York University, New York, NY).

Gordon, Bruce, The History Makers, July 14, 2013. https://www.thehistorymakers.org/biography/bruce-gordon.

Grandy, Alicia A., James M. Diefendorr, and Deborah E. Rupp, eds. *Emotional Labor in the 21st Century: Diverse Perspectives on Emotion Regulation at Work*. New York: Routledge, 2013.

Green, James. *The World of the Workers: Labor in 20th Century America*. New York: Hill and Wang, 1980.

Green, Venus. "Flawed Remedies: EEOC, AT&T, and Sears Outcomes Reconsidered," *Black Women Gender + Families* 6, no.1 (spring 2012): 43–70.

Green, Venus. *Race on the Line: Gender, Labor, and Technology in the Bell System, 1880–1980*. Durham, NC: Duke University Press, 2001.

Greenbaum, Joan. *Windows on the Workplace: Computers, Jobs, and the Organization of Office Work in the Late Twentieth Century*. New York: Monthly Review Press, 1995.

Greenhouse, Steven. *Beaten Down: Worked Up: The Past, Present, and Future of American Labor.* New York: Knopf, 2019.

Greenhouse, Steven. *The Big Squeeze: Tough Times for the American Worker.* New York: Knopf, 2008.

Gutman, Herbert G. *Power and Culture: Essays on the American Working Class.* New York: Pantheon Books, 1987.

Gutman, Herbert G. *Work, Power, and Culture in Industrializing America: Essays in American Working-Class and Social History*. New York: Knopf, 1976.

Hacker, Sally. "Sex Stratification, Technology and Organizational Change: A Longitudinal Case Study of AT&T." *Social Problems* 26 (June 1979): 539–70.

Hafiz, Hifa. "Rethinking Breakups." *Duke Law Journal* 71 (2022): 1491–1603.

Hammer, Michael. "Reengineering Work: Don't Automate, Obliterate." *Harvard Business Review* 68, no. 4 (July-August 1990): 104–12

Hammer, Michael, and James Champy. *Reengineering the Corporations: A Manifesto for Business Revolution*. New York: Harper Collins, 1993.

Harrison, Maryellen R., and Bennett Harrison. "Unions, Technology, and Labor-Management Cooperation." In *Unions and Economic Competitiveness,* edited by Lawrence Mishel and Paula Voos, 247–282. Armonk, NY: M.E. Sharpe, 1992.

Hartmann, Heidi I. *Computer Chips and Paper Clips: Technology and Women's Employment*. Vol. 2. Washington, DC: National Academy Press, 1987.

Hartmann, Heidi I., Robert E. Kraut, and Louise A. Tilly. *Computer Chips and Paper Clips: Technology and Women's Employment*. Vol. 1. Washington, DC: National Academy Press, 1986.

Hartmann, Heidi I., Robert T. Michael, and Brigid O'Farrell. *Pay Equity: Empirical Inquiries.* Washington, DC: National Research Council, 1989.

Harvey, David. *A Brief History of Neoliberalism*. New York: Oxford University Press, 2015.

Heckscher, Charles, Michael Maccoby, Rafael Ramirez, and Pierre-Eric Tixier. *Agents of Change: Crossing the Post-Industrial Divide*. New York: Oxford University Press, 2003.

Heckscher, Charles, Sue Schurman, Adrienne Eaton, and Beth Craig. "Work Place of the Future: A Research Report." Rutgers, NJ: Rutgers University Department of Labor Studies and Employment Relations, May 15, 1997.

Hechanova-Alampay, Ma. Regina. *1-800-philipines: Understanding and Managing the Filipino Call Center Worker*. Quezon City, Philippines: Institute of Philippine Culture, 2010.

Herr, Lois Kathryn. *Women, Power, and AT&T: Winning Rights in the Workplace*. Boston: Northeastern University Press, 2003.

Hiatt, John P., and Lee W. Jackson. "Union Survival Strategies for the Twenty-First Century." *Journal of Labor Research* 18, no. 4 (Fall 1997): 487–501.

Hilton, Margaret, Gretchen Kolsrud, Peter D. Blair, and Audrey B. Buyrn. *Pulling Together for Productivity: A Union-Management Initiative at US West, Inc*. Washington, DC: U.S. Office of Technology Assessment, OTA-ITE-583, September 1993.

Hirsch, Barry T., and David MacPherson. "Union Membership, Coverage, Density and Employment from the CPS," http://unionstats.com.

Hochschild, Arlie. *The Managed Heart: Commercialization of Human Feeling*. Berkeley: University of California Press, 1983.

Holman, David, Rosemary Batt, and Ursula Holtgrewe. *The Global Call Center Report: International Perspectives on Management and Employment: Report of the Global Call Center Network*. Ithaca, NY, 2007.

Holman, David, Stephen Frenkel, Old Sorensen, and Stephen Wood. "Work Design Variation and Outcomes in Call Centers: Strategic Choice and Institutional Explanations." *Industrial and Labor Relations Review* 62, no. 4 (2009): 510–32.

Horwitz, Robert Britt. *The Irony of Regulatory Reform: The Deregulation of American Telecommunications*. New York: Oxford University Press, 1989.

Hower, Joseph E. "You've Come a Long Way—Maybe": Working Women, Comparable Worth, and the Transformation of the American Labor Movement, 1964–1989." *Journal of American History*, 107, no. 3 (December 2020), 658–84.

Huber, Peter W., Michael K. Kellogg, and John Thorn. *The Geodesic Network: 1993 Report on Competition in the Telephone Industry*. Washington, DC: Geodesic Company, 1992.

Jacoby, Sanford M. *Employing Bureaucracy: Managers, Unions, and the Transformation of Work in the 20th Century*. Mahwah, NJ: Lawrence Erlbaum, 2004.

Jacoby, Sanford M. *Labor in the Age of Finance: Pensions, Politics, and Corporations from Deindustrialization to Dodd-Frank*. Princeton, NJ: Princeton University Press, 2021.

Jamakaya. *Like Our Sisters Before Us: Women of Wisconsin Labor.* Wisconsin Labor History Society, 1998, 23–34.
James, Daniel. *Dona Marie's Story.* Durham, NC: Duke University Press, 2000.
Jensen, Michael C., and Kevin J. Murphy. "CEO Incentives—It's Not How Much You Pay, But How." *Harvard Business Review* 68, no. 3 (May-June 1990): 138–49.
John, Richard. *Network Nation: Inventing American Telecommunications.* Cambridge, MA: Harvard University Press, 2010.
Juravich, Tom. *At the Altar of the Bottom Line: The Degradation of Work in the 21st Century.* Amherst: University of Massachusetts Press, 2009.
Kahaner, Larry. *On the Line: The Men of MCI—Who Took on AT&T, Risked Everything, and Won!* New York: Warner Books, 1986.
Kahn, Alfred. *The Economics of Regulation: Principles and Institutions.* Vol. 1. New York: Wiley, 1970.
Kalleberg, Arne. *Good Jobs/Bad Jobs: The Rise of Polarized and Precarious Employment Systems in the United States, 1970s to 2000s.* New York: Russell Sage Foundation, 2011.
Kanigel, Robert. *The One Best Way.* Cambridge, MA: MIT Press, 2005.
Karasek Jr., Robert. "Job Demands, Job Decision Latitude, and Mental Strain: Implications for Job Redesign." *Administrative Science Quarterly* 24, no. 2 (June 1979): 285–308.
Karl, Frank. "A Panorama of Collective Bargaining: The Communications Workers of America and the American Telephone and Telegraph Company, 1938–1989." MA Thesis, State University of New York, Empire State Colleges, 1991.
Katz, Harry. *Shifting Gears: Changing Labor Relations in the US Automobile Industry.* Cambridge, MA: MIT Press, 1985.
Katz, Harry, ed. *Telecommunications: Restructuring Work and Employment Relations Worldwide.* New York: ILR Press, 1997.
Katz, Harry, Rosemary Batt, and Jeffrey H. Keefe. "The Revitalization of the CWA: Integrating Collective Bargaining, Political Action, and Organizing." *Industrial and Labor Relations Review* 56, no. 4 (2003): 573–89.
Katz, Harry, Thomas A. Kochan, and Alexander J.S. Colvin. *Labor Relations in a Globalizing World.* Ithaca, NY: ILR Press, 2015.
Kaufman, Bruce E., and Bennett, James, T., eds. *What Do Unions Do? A Twenty-Year Perspective.* New Brunswick, NJ: Transaction Publishers, 2007.
Keefe, Jeffrey. "Is Digital Technology Reshaping Employment Systems in U.S. Telecommunications Network Services?" *Industrial and Labor Relations Review* 63, no. 1 (2009): 42–59.
Keefe, Jeffrey. *Racing to the Bottom: How Antiquated Public Policy Is Destroying the Best Jobs in Telecommunications.* Washington, DC: Economic Policy Institute, 2005.
Keefe, Jeffrey, and Rosemary Batt. "Telecommunications: Collective Bargaining in an Era of Industrial Reconsolidation." In *Collective Bargaining in the*

Private Sector, edited by Paul R. Clark, John T. Delaney, and Ann C. Frost, 263–310. Champaign, IL: Industrial Relations Research Association, 2002.

Keefe, Jeffrey, and Karen Boroff. "Telecommunications Labor-Management Relations: One Decade after the AT&T Divestiture." In *Contemporary Collective Bargaining in the Private Sector*, edited by Paula Voos. Madison, WI: Industrial Relations Research Association, 1994.

Keefe, Jeffrey, and George Kohl. "Technological Change and Labor's Interests." *Workplace Topics* 3, no. 1 (September 1993): 1–32, Washington, DC: AFL-CIO Department of Economic Research .

Kessler-Harris, Alice. *Gendering Labor History*. Urbana: University of Illinois Press, 2007.

Kessler-Harris, Alice. *In Pursuit of Equity: Women, Men, and the Quest for Economic Citizenship in Twentieth-Century America*. New York: Oxford University Press, 2003.

Kessler-Harris, Alice. *Out to Work*. New York: Oxford University Press, 1982.

Kessler-Harris, Alice. "Where Are the Organized Women Workers?" *Feminist Review* 3 (Fall 1975): 92–105.

Kleinfield, Sonny. *The Biggest Company on Earth*. New York: Holt, Rinehart and Winston, 1981.

Kloepfer, Joan H. *Inside Sprint Corp*. Alexandra, VA: Telecom Publishing, 1993.

Kochan, Thomas A., Harry C. Katz, and Robert B. McKersie. *The Transformation of American Industrial Relations*. New York: Basic Books, 1986.

Korczynski, Marek, and Cameron Lynn Macdonald, eds. *Service Work: Critical Perspectives*. New York: Routledge, 2009.

Kotz, David M. *The Rise and Fall of Neoliberal Capitalism*. Cambridge, MA: Harvard University Press, 1971.

Krippner, Greta R. *Capitalizing on Crisis: The Political Origins of the Rise of Finance*. Cambridge, MA: Harvard University Press, 2011.

Krugman, Paul. *Peddling Prosperity: Economic Sense and Nonsense in an Age of Diminished Expectations*. New York: Norton, 1995.

Langer, Elinor. "Inside the New York Telephone Company." *New York Review of Books*, March 12, 1970, 16–24.

Langer, Elinor. "The Women of the Telephone Company." *New York Review of Books*, March 26, 1970, 14–22.

Lauer, Josh, and Kenneth Lipartito, eds. *Surveillance Capitalism in America*. Philadelphia: University of Pennsylvania Press, 2021.

Lazonick, William. "The Financialized Corporation and American Income Inequality." *Perspectives on Work* 19, no. 1 (2015): 30–33, 83–87.

Lazonick, William. "Profits without Prosperity." *Harvard Business Review* (September 2014): 3–11.

Lazonick, William, and William Shin. *Predatory Value Extraction: How the Looting of the Business Corporation Became the U.S. Norm and How Sustainable Prosperity Can Be Restored*. Oxford: Oxford University Press, 2020.

Lerner, Stephen, Jill Hurst, and Glenn Adler. "Fighting and Winning in the Outsourced Economy: Justice for Janitors at the University of Miami." In

The Gloves Off Economy: Workplace Standards at the Bottom of the Labor Market, edited by Annette Bernhardt, Heather Boushey, Laura Dresser, and Chris Tilley. Champaign, IL: Labor and Employment Research Association, 2008.

Levine, David L., and Laura D'Andrea Tyson. "Participation, Productivity, and the Firm's Environment." In *Paying for Productivity,* edited by Alan Blinder. Washington, DC: Brookings Institution, 1990.

Levinson, Marc. *An Extraordinary Time: The End of the Postwar Boom and the Return of the Ordinary Economy*. New York: Basic Books, 2016.

Levitt, Martin Jay, and Terry Conrow. *Confessions of a Union Buster.* New York: Crown Publishers, 1993.

Levy, Karen. *Data Driven: Truckers, Technology, and the New Workplace Surveillance*. Princeton, NJ: Princeton University Press, 2022.

Lichtenstein, Nelson. "From Corporatism to Collective Bargaining: Organized Labor and the Eclipse of Social Democracy in the Postwar Era." In *The Rise and Fall of the New Deal Order,* edited by Steven Fraser and Gary Gerstle. Princeton, NJ: Princeton University Press, 1989.

Lichtenstein, Nelson. *The State of the Union: A Century of American Labor*. Princeton, NJ: Princeton University Press, 2002.

Lichtenstein, Nelson, and Judith Stein. *A Fabulous Failure: The Clinton Presidency and the Transformation of American Capitalism*. Princeton, NJ: Princeton University Press, 2023.

Lipartito, Ken. *The Bell System and Regional Business: The Telephone in the South, 1877–1920*. Baltimore, MD: Johns Hopkins University, 1989.

Lipartito, Ken. "When Women Were Switches: Technology, Work, and Gender in the Telephone Industry, 1890–1920." *The American Historical Review* 99, no. 4 (Oct. 1994): 1075–11.

Lloyd, Caroline, Claudia Weinkopf, and Rosemary Batt. "Restructuring Customer Services: Labor Market Institutions and Call Center Workers in Europe and the United States." In *Low Wage Work in a Wealthy World,* edited by Jerome Gautie and John Schmitt. New York: Russell Sage Foundation, 2009.

Lowe, Sarah. "The First American Case under the North American Agreement for Labor Cooperation." *University of Miami Law Review* 481 (January 1997): 481–510.

Lynd, Stoughton. *We Are All Leaders: The Alternative Unionism of the Early 1930s*. Urbana: University of Illinois Press, 1996.

MacAvoy, Paul. *The Regulated Industries and the Economy*. New York: Norton, 1979.

Maccoby, Michael. "Helping Labor and Management Set Up a Quality-of-Work Life Program," *Monthly Labor Review*, March 1984, 28–32.

Macdonald, Cameron Lynne, and Carmen Sirianni, eds. *Working in the Service Society*. Philadelphia: Temple University Press, 1996.

MacLean, Nancy. *Freedom Is Not Enough: The Opening of the American Workplace*. Cambridge, MA: Harvard University Press, 2006.

Madland, David. *Re-Union: How Bold Labor Reforms Can Repair, Revitalize, and Reunite the United States.* Ithaca, NY: Cornell University Press, 2021.

Marshall, Ray. "Unions, Work Organization, and Economic Performance." In *Unions and Economic Competitiveness,* edited by Lawrence Mishel and Paula Voos. Armonk, NY: M.E. Sharpe, 1992.

Marx, Karl. *The Eighteenth Brumaire of Louise Bonaparte.* In *Karl Marx: A Reader,* edited by John Elster. New York: Press Syndicate of the University of Cambridge, 1986.

Mateescu, Alexandra, and Aiha Nguyen. "Algorithmic Management in the Workplace." *Data & Society,* February 2019.

McCartin, Joseph A. "'As Long as There Survives': Contemplating the Wagner Act after Eighty Years." *Labor: Studies in Working Class History* 14, no. 2 (May 2017): 21–42.

McCartin, Joseph A. *Collision Course: Ronald Reagan, the Air Traffic Controllers, and the Strike That Changed America.* New York: Oxford University Press, 2011.

McCartin, Joseph A. "Context Matters Most: A Response to Joe Burns." *Labor Studies Journal* 37, no. 4 (2013): 349–52.

McCartin, Joseph A. "Solvents of Solidarity: Political Economy, Collective Action, and the Crisis of Organized Labor, 1968–2005." In *Rethinking U.S. Labor History: Essays on the Working-Class Experience 1756–2009,* edited by Donna T. Haverty-Stacke and Daniel J. Walkowitz. New York: Continuum, 2010.

McCaughey, Ewan. "The Codetermination Bargains: The History of German Corporate Law and Labour Law." *London School of Economics Law, Society, and Economy Working Papers,* March 31, 2021 (last revised). https://papers.ssrn.com/sol3/papers.cfm?abstract_id=2579932.

McCraw, Thomas K. *Prophets of Regulation: Charles Francis Adams, Louis D. Brandeis, James M. Landis, Alfred E. Kahn.* Cambridge, MA: Harvard University Press, 1984.

McNicholas, Celine, Margaret Poydock, Julia Wolfe, Ben Zipperer, Gordon Lafer, and Lola Loustaunau. *Unlawful: U.S. Employers Are Charged with Violating Law in 41.5% of All Union Election Campaigns.* Washington, DC: Economic Policy Institute, December 11, 2019.

Metzgar, Jack. *Striking Steel: Solidarity Remembered.* Philadelphia: Temple University Press, 2000.

Meyer, Stephen III. *The Five Dollar Day: Labor Management and Social Control in the Ford Motor Company.* Albany: State University of New York Press, 1981.

Miller, Jeffrey. "The Bossless Office: Unique Arizona Experiment Proves Workers Can Run the Show." *CWA News,* February 1984.

Milkman, Ruth. *Gender at Work: The Dynamics of Job Segregation by Sex During World War II.* Urbana: University of Illinois Press, 1987.

Milkman, Ruth, ed. *Women, Work, and Protest: A Century of U.S. Women's Labor History.* Boston: Routledge and Kegan Paul, 1985.

Milkman, Ruth, and Kim Voss, eds. *Rebuilding Labor: Organizing and Organizers in the New Labor Movement*. Ithaca, NY: ILR Press, 2004.

Miscimarra, Philip A., and Kenneth D. Schwartz. "Frozen in Time—The NLRB, Outsourcing, and Management Rights." *Journal of Labor Research* 18, no. 4 (Fall 1997): 561–80.

Moberg, David. "Rattling the Golden Chains: Conflict and Consciousness of Auto Workers," PhD diss., University of Chicago, March 1978.

Montgomery, David. *Worker's Control in America: Studies in the History of Work, Technology, and Labor Struggles*. New York: Cambridge University Press, 1987.

Moody, Kim. *An Injury to All: The Decline of American Unionism*. New York: Verso, 1988.

Moss, Philip, Harold Salzman, and Chris Tilly. "Under Construction: The Continuing Evolution of Job Structures in Call Centers." *Industrial Relations* 47, no. 2 (2008): 173–208.

Murphy, Ryan Patrick. *Deregulating Desire: Flight Attendant Activism, Family Politics, and Workplace Justice*. Philadelphia: Temple University Press, 2016.

Nadasen, Primilla. *Household Workers Unite: The Untold Story of African American Women Who Built a Movement*. Boston: Beacon Press, 2015.

Nelson, Daniel. *Frederick W. Taylor and the Rise of Scientific Management*. Madison: University of Wisconsin Press, 1980.

Neuchterlein, Jonathon E., and Philip J. Weiser. *Digital Crossroads: Telecommunications Law and Policy in the Internet Age*. Cambridge, MA: MIT Press, 2013.

Nielsen, Georgia Panter. *From Sky Girl to Flight Attendant: Women and the Making of a Union*. Ithaca, NY: ILR Press, 1982.

Nissen, Bruce, ed. *Unions in a Globalized Environment: Changing Borders, Organizational Boundaries, and Social Roles*. Armonk, NY: M.E. Sharpe, 2002.

Noble, David. *Forces of Production: A Social History of Industrial Automation*. New York: Knopf, 1984.

Northrup, Herbert R., and John A. Larson. *The Impact of the AT&T-EEO Consent Decree*. Philadelphia: University of Pennsylvania Press, 1979.

Norwood, Stephen. *Labor's Flaming Youth: Telephone Operators and Worker Militancy, 1878–1923*. Urbana: University of Illinois, 1990.

Office of Technology Assessment. *The Electronic Supervisor: New Technology, New Tensions*. OTA-CIT-333. Washington, DC: Government Printing Office, 1987.

Pacific Bell. *Pacific Bell's Response to the Intelligent Network Task Force Report*. 1988.

Padios, Jan Maghinay. "Listening Between the Lines: Culture, Difference, and Immaterial Labor in the Philippine Call Center Industry." PhD diss., New York University, 2012.

Page, Arthur. *The Bell Telephone System*. New York: Harper, 1941.

Palley, Thomas L. "Financialization: What It Is and Why It Matters." Annandale-on-Hudson, NY: Levey Economics Institute of Bard College, 2007.

Parker, Mike. *Inside the Circle: A Union Guide to QWL*. Boston: South End Press, 1985.

Parker, Mike. "Industrial Relations Myth and Shop-Floor Reality: The 'Team Concept' in the Auto Industry." In *Industrial Democracy in America: The Ambiguous Promise*, edited by Nelson Lichtenstein and Howell John Harris. Cambridge University Press, 249–74.

Parker, Mike. *Working Smart: A Union Guide to Participation Programs and Reengineering*. Detroit: Labor Education and Research Project, 1994.

Parker, Mike, and Jane Slaughter. *Choosing Sides: Unions and the Team Concept*. Boston: South End Press, 1988.

Parker, Traci. *Department Stores and the Black Freedom Movement: Workers, Consumers, and Civil Rights from the 1930s to the 1980s*. Chapel Hill: University of North Carolina Press, 2019.

Pattee, Jon. "Sprint and the Shutdown of La Conexion Familiar: A Union Hating Multinational Finds Nowhere to Run." *Labor Research Review* 1, no. 23 (1995): 13–21.

Perlman, Selig. *A History of Trade Unionism in the United States*. New York: Macmillan, 1922.

Perry, L.J., and Patrick J. Wilson, "Trends in Work Stoppages: A Global Perspective." Working Paper No. 47, Policy Integration Department, Statistical Analysis Unit. Geneva: International Labor Organization, 2004. http://papers.ssrn.com/sol3/papers.cfm?abstract_id=908483.

Pew Research Center. *Mobile Technology and Home Broadband, 2021*, June 3, 2021.

Piore, Michael, and Charles Sabel. *The Second Industrial Divide: Possibilities for Prosperity*. New York: Basic Books, 1984.

Portelli, Alessandro. *The Death of Luigi Trastulli and Other Stories: Form and Meaning in Oral History*. Albany: State University of New York Press, 1991.

Rechenbach, Jeff, and Larry Cohen. "Union Global Alliances at Multinational Corporations: A Case Study of the Ameritech Alliance." In *Unions and Workplace Reorganization*, edited by Bruce Nissen. Detroit: Wayne State University Press, 1997.

Reich, Robert. *The Next American Frontier*. New York: Times Books, 1983.

Reich, Robert. *Supercapitalism: The Transformation of Business, Democracy, and Everyday Life*. New York: Knopf, 2007.

Reich, Robert. *The Wealth of Nations: Preparing Ourselves for 21st Century Capitalism*. New York: Vintage Books, 1992.

Reiman, Donald J., and Heidi I. Hartman, eds. *Women, Work, and Wages: Equal Pay for Jobs of Equal Value*. Washington, DC: National Academy Press, 1981.

Resnikoff, Jason. *Labor's End: How the Promise of Automation Degraded Work*. Urbana: University of Illinois Press, 2021.

Reyes, Victoria. "Ethnographic Toolkit: Strategic Positionality and Researchers' Visible and Invisible Tools in Field Research." *Ethnography* 21, no. 2 (2020): 220–40.
Rifkin, Jeremy, and Randy Barber. *The North Will Rise Again: Pensions, Politics, and Power in the 1980s*. Boston: Beacon Press, 1978.
Rodgers, Daniel T. *Age of Fracture*. Cambridge, MA: Harvard University Press, 2011.
Roscoe, Jules. "Amazon Workers Are Still Peeing in Bottles." *Motherboard*, November 2, 2022.
Roseman, Herman G., and Irwin M. Stelzer. *Economic Problems of Regulated Competition*. New York: National Economic Research Associates, 1975.
Rothschild, Joan, ed. *Machina Ex Dea: Feminist Perspectives on Technology*. New York: Pergamon Press, 1983.
Russell, Bob. *Smiling Down the Line: Info-Service Work in the Global Economy*. Toronto: University of Toronto Press, 2009.
Sabel, Charles. *Work and Politics: The Division of Labor in Industry*. New York: Cambridge University Press, 1982.
Schacht, John. *The Making of Telephone Unionism 1920–1947*. New Brunswick, NJ: Rutgers University Press, 1985.
Schacht, John. "Toward Industrial Unionism: Bell Telephone Workers and Company Unions, 1919–1937." *Labor History* 16, no. 1 (1975): 5–36.
Schiller, Dan. *Telematics and Government*. Norwood, NJ: Ablex, 1982.
Schlesinger, Leonard A., Davis Dyer, Thomas M. Clough, and Diane Landau. *Chronicles of Corporate Change: Management Lessons from AT&T and Its Offspring*. Lexington, MA: Lexington Books, 1987.
Schmitt, John, and Jori Kandra. *Decades of Slow Wage Growth for Telecommunication Workers*. Washington, DC: Economic Policy Institute, 2020.
Schmitt, John, and Kris Warner. *The Changing Face of Labor 1983–2008*. Washington, DC: Center for Economic and Policy Research, 2009.
Scott, Joan. "Gender: A Useful Category of Historical Analysis." *The American Historical Review* 91, no. 5 (December 1986): 1053–75.
Secretariat of the Commission for Labor Cooperation. "Plant Closings and Labor Rights: The Effects of Sudden Plant Closings on Freedom of Association and the Right to Organize in Canada, Mexico and the United States." June 9, 1997.
Semel, Gabrielle. *The Cablevision War: How a Majority Black Workforce and a Union Willing to Fight Won against a Billionaire* (forthcoming).
Shaiken, Harley. *Work Transformed: Automation and Labor in the Computer Age*. Lexington, MA: DC Heath, 1986.
Sheth, Jagdish N., and David A. Heffner. *Voice with a Smile: True Stories of Telephone Operators*. Barrington, IL: Perq Publications, 1991.
Slichter, Sumner, James J. Healy, and E. Robert Livernash. *The Impact of Collective Bargaining on Management*. Washington, DC: Brookings Institution, 1960.

Smith, George David. *The Anatomy of Business Strategy: Bell, Western Electric, and the Origins of the American Telephone Industry*. Baltimore, MD: Johns Hopkins University Press, 1985.

Sneiderman, Marilyn, and Stephen Lerner. "Making Hope and History Rhyme: A New Worker Movement from the Shell of the Old." *New Labor Forum*, December 13, 2022.

Spalter-Roth, Roberta, and Heidi Hartmann. "Women in Telecommunications: Exception to the Rule of Low Pay for Women's Work." Washington, DC: Institute for Women's Policy Research, 1992.

Stedman Jones, Daniel. *Master of the Universe: Hayek, Friedman, and the Birth of Neoliberal Politics*. Princeton, NJ: Princeton University Press, 2012.

Stein, Judith. *Pivotal Decade: How the U.S. Traded Factories for Finance in the 1970s*. New Haven, CT: Yale University Press, 2010.

Stevens, Andrew J.R. *Call Centers and the Global Division of Labor: A Political Economy of Post-Industrial Employment and Union Organizing*. New York: Routledge, 2014.

Stigler, George. "The Theory of Economic Regulation." *The Bell Journal of Economics & Management Science* 2, no. 1 (1971): 3–21.

Stiglitz, Joseph. *Freefall: America, Free Markets, and the Sinking World Economy*. New York: Norton, 2010.

Stockford, Marjorie A. *The Bellwomen: The Story of the Landmark AT&T Sex Discrimination Case*. New Brunswick, NJ: Rutgers University Press, 2004.

Stone, Alan. *Wrong Number: The Breakup of AT&T*. New York: Basic Books, 1989.

StrategyR, Global Industry Analysts, Inc. https://www.strategyr.com/market-report-call-centers-forecasts-global-industry-analysts-inc.asp (viewed June 9, 2021).

Strauss, George. "Worker Participation—Some Under-Considered Issues." *Industrial Relations* 45, no. 4 (September 2006): 778–803.

Strom, Sharon Hartman. *Beyond the Typewriter: Gender, Class, and the Origins of Modern American Office Work, 1900–1930*. Urbana: University of Illinois Press, 1992.

Sweeney, John J., and Karen Nussbaum. *Solutions for the New Work Force: Policies for a New Social Contract*. Cabin John, MD: Seven Locks Press, 1989.

Sugrue, Thomas. *The Origins of the Urban Crisis: Race and Inequality in Postwar Detroit*. Princeton, NJ: Princeton University Press, 1996.

Taylor, Frederick. *The Principles of Scientific Management*. New York: Harper, 1911.

Taylor, Frederick, Gareth Mulvey, Jeff Hyman, and Peter Bain. "Work Organization, Control and Experience of Work in Call Centres." *Work, Employment, and Society* 16, no. 1 (2002): 133–50.

Taylor, Phillip, and Peter Bain. "An Assembly-Line in the Head: The Call Center Labor Process," *Industrial Relations Journal* 30, no. 2 (1999): 101–17.

Tedlow, Richard S., and Alfred D. Chandler Jr. *The Coming of Managerial Capitalism: A Casebook on the History of American Economic Institutions.* Homewood, IL: R.D. Irwin, 1985.

Temin, Peter, and Louis Galambos. *The Fall of the Bell System.* New York: Cambridge University Press, 1987.

Thompson, E.P. *The Making of the English Working Class.* New York: Pantheon Books, 1964.

Tunstall, W. Brooke. *Disconnecting Parties: Managing the Bell System Break-Up: An Inside View.* New York: McGraw-Hill, 1985.

Turk, Katherine. *Equality on Trial: Gender and Rights in the Modern American Workplace.* Philadelphia: University of Pennsylvania Press, 2016.

Turner, Lowell, Harry Katz, and Richard Hurd, eds. *Rekindling the Movement: Labor's Quest for Relevance in the 21st Century.* Ithaca, NY: Cornell University Press, 2001.

Unger, Laura. Speech for Workplace of the Future Forum, February 7, 1995.

U.S. Department of Labor. "Quality of Work Life: AT&T and CWA Examine Process after Three Years." 1985.

U.S. Department of Labor and U.S. Department of Commerce. U.S. Commission on the Future of Worker-Management Relations. *Fact-Finding Report,* May 1994.

U.S. Department of Labor. U.S. Commission on the Future of Worker-Management Relations. *Report and Recommendations,* December 1994.

Vallas, Steven Peter. *Power in the Workplace: The Politics of Production at AT&T.* Albany: SUNY Press, 1993.

Vietor, Richard H.K. *Contrived Competition: Regulation and Deregulation in America.* Cambridge, MA: Harvard University Press, 1994.

Vogel, David. *Fluctuating Fortunes.* New York: Basic Books, 1989.

Voss, Kim, and Rachel Sherman. "Breaking the Iron Law of Oligarchy: Union Revitalization and the American Labor Movement." *American Journal of Sociology* 106, no. 2 (September 2000): 303–49.

Wallace, Phyllis, ed. *Equal Employment Opportunity and the AT&T Case.* Cambridge, MA: MIT Press, 1976.

Walton, Richard E. "From Control to Commitment: Transforming Work Force Management in the United States." Prepared for the Harvard Business School's 75th Anniversary Colloquium on Technology and Productivity, March 27–29, 1984.

Weil, David. *The Fissured Workplace: Why Work Became So Bad for So Many and What Can Be Done to Improve It.* Cambridge, MA: Harvard University Press, 2014.

Weiler, Paul. *Governing the Workplace: The Future of Labor and Employment Law.* Cambridge, MA: Harvard University Press, 1990.

Wells, Donald M. *Empty Promises: Quality of Working Life Programs and the Labor Movement.* New York: Monthly Review Press, 1987.

Whetzell, John. "Absence Control in the Telephone Industry." Communications Workers of America, January 4, 1974.
Williamson, Oliver. *The Economic Institutions of Capitalism*. New York: Free Press, 1985.
Winant, Gabriel. *The Next Shift: The Fall of Industry and the Rise of Health Care in Rust Belt America*. Cambridge, MA: Harvard University Press, 2021.
Windham, Lane. *Knocking on Labor's Door: Union Organizing in the 1970s and the Roots of a New Economic Divide*. Chapel Hill: University of North Carolina Press, 2017.
Woodcock, Jamie. *Working the Phones: Control and Resistance in Call Centres*. London: Pluto Press, 2017.
Wright, Barbara Drygulski, ed. *Women, Work, and Technology*. Ann Arbor: University of Michigan Press, 1987.
Zickuhr, Kathryn. *Workplace Surveillance Is Becoming the New Normal for U.S. Workers*. Washington, DC: Washington Center for Equitable Growth, August 2021.
Zuboff, Shoshana. *In the Age of the Smart Machine: The Future of Work and Power*. New York: Basic Books, 1988.
Zuboff, Shoshana. *The Age of Surveillance Capitalism*. New York: Public Affairs, 2019.

Index

Note: Page numbers in *italics* denote tables and figures.

DEBBIE J. GOLDMAN is the former Research Director and Telecommunications Policy Director with the Communications Workers of America.

The University of Illinois Press
is a founding member of the
Association of University Presses.

University of Illinois Press
1325 South Oak Street Champaign,
IL 61820-6903
www.press.uillinois.edu